T0342111

Smart Grid Communication Infrastructures

Smart Grid Communication Infrastructures

Big Data, Cloud Computing, and Security

Feng Ye
University of Dayton
Dayton, Ohio

Yi Qian
University of Nebraska-Lincoln
Omaha, Nebraska

Rose Qingyang Hu
Utah State University
Logan, Utah

The right of Feng Ye, Yi Qian & Rose Qingyang Hu to be identified as the authors of this work has been asserted in accordance with law.

Registered Offices
John Wiley & Sons, Inc., 111 River Street, Hoboken, NJ 07030, USA
John Wiley & Sons Ltd, The Atrium, Southern Gate, Chichester, West Sussex, PO19 8SQ, UK

Editorial Office
The Atrium, Southern Gate, Chichester, West Sussex, PO19 8SQ, UK

For details of our global editorial offices, customer services, and more information about Wiley products visit us at www.wiley.com.

Wiley also publishes its books in a variety of electronic formats and by print-on-demand. Some content that appears in standard print versions of this book may not be available in other formats.

Library of Congress Cataloging-in-Publication Data
Names: Ye, Feng, 1989- author. | Qian, Yi, 1962- author. | Hu, Rose Qingyang, author.
Title: Smart grid communication infrastructures : big data, cloud computing, and security / by Feng Ye, Yi Qian, and Rose Qingyang Hu.
Description: Hoboken, NJ : John Wiley & Sons, 2018. | Includes bibliographical references and index. |
Identifiers: LCCN 2018001007 (print) | LCCN 2018012065 (ebook) | ISBN 9781119240181 (pdf) | ISBN 9781119240167 (epub) | ISBN 9781119240150 (cloth)
Subjects: LCSH: Smart power grids–Communication systems. | Smart power grids–Security measures.
Classification: LCC TK3105 (ebook) | LCC TK3105 .Y44 2018 (print) | DDC 621.31–dc23
LC record available at https://lccn.loc.gov/2018001007

Cover Design: Wiley
Cover Image: © DuKai photographer/Getty Images

Set in 10/12pt WarnockPro by SPi Global, Chennai, India

Printed in Singapore by C.O.S. Printers Pte Ltd

10 9 8 7 6 5 4 3 2 1

Contents

1

Background of the Smart Grid

The world has been rapidly developing higher efficiency in many aspects. As one of these important aspects, power grids in many countries have been evolving from traditional power grids into *smart grids* in recent years. What is a smart grid? In this chapter, we will introduce the background of the smart grid, including its motivations, communication architecture, applications, and requirements.

1.1 Motivations and Objectives of the Smart Grid

The traditional power grid, or just "the grid," is a network of transmission lines, substations, transformers, and more that delivers electricity from the power plant to consumers (e.g. your home, business, etc.). Our current power grid was built over a century ago. To move forward, we need a new kind of power grid that can automate and manage the increasing need for and complexity of delivering electricity. You may have heard of this new kind of power grid, which is called the *smart grid*. The smart grid is a revolutionary upgrade to the traditional power grid that adds communication capabilities, intelligence, and modern control [1–10]. The US Department of Energy (DoE), Office of Electricity Delivery & Energy Reliability has listed the benefits associated with the smart grid as follows [11]:

- More efficient transmission of electricity.
- Quicker restoration of electricity after power disturbances.
- Reduced operations and management costs for utilities and ultimately lower power costs for consumers.

Smart Grid Communication Infrastructures: Big Data, Cloud Computing, and Security,
First Edition. Feng Ye, Yi Qian, and Rose Qingyang Hu.
© 2018 John Wiley & Sons Ltd. Published 2018 by John Wiley & Sons Ltd.

- Reduced peak demand, which will also help to reduce electricity rates.
- Increased integration of large-scale renewable energy systems.
- Better integration of customer-owner power generation systems, including renewable energy systems.
- Improved security.

We further summarize the benefits into three major motivations for the smart grid: 1) To better adapt renewable energy resources, 2) to increase the efficiency of grid operations and to reduce losses, and 3) to improve system reliability and security.

1.1.1 Better Renewable Energy Resource Adaption

In the current power grid, the majority of the power generation stations are based on fossil fuels (e.g. coal, natural gas, petroleum, etc.) [12]. While such power stations have supported civilization for over a century, we have come to realize the shortcomings and disadvantages of such power supplies. Since fossil fuel is limited resource, its ever-decreasing supply will ultimately affect the power stations. Moreover, the emission of greenhouse gases from those power stations has gradually contributed to global climate change, especially global warming. Therefore, cleaner energy resources and renewable energy resources have been deployed to the grid in the past few decades. Some examples of existing power stations using cleaner energy resources are hydropower, geothermal, etc.

The smart grid will include both central and distributed generation sources with a mix of dispatchable and nondispatchable resources, as illustrated in Figure 1.1. Despite their obvious benefits, renewable sources other than hydropower provide only about 5% of the electricity supply for our grid. What's holding us back is not the amount of electricity we can generate from those renewable resources but the power grid itself. Due to the remote locations of most renewable resources, extra infrastructure is required to deliver electricity to consumers, who are mostly in urban areas. Moreover, the current power grid has difficulty accommodating renewable sources of power due to their unpredictability. The smart grid will have much better control systems to manage those renewable energy resources to supplement existing fossil-fuel power stations. For example, the smart grid will give grid operators new tools to reduce power demand

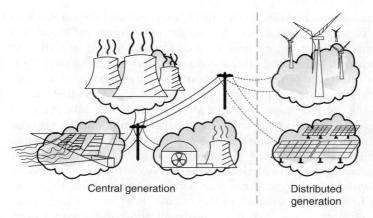

Central generation | Distributed generation

Figure 1.1 Dispatchable renewable resources.

quickly when the level of wind or solar power dips, and it will have more energy storage capabilities to absorb excess wind and solar power when it isn't needed, then to release that energy when the level of wind and solar power dips.

1.1.2 Grid Operation Efficiency Advancement

The smart grid is able to increase the efficiency of the grid operations while reducing losses because of its advanced monitoring and controlling systems. In the current power grid, the pricing of electricity is based on peak and off-peak demand. The price of electricity during peak hours is higher than during off-peak hours. For some consumers, the higher price may be a few times more than the lower one. The uneven pricing of electricity is due to the dynamic cost of power generation, especially in fossil-fuel-based power stations. It is costlier per unit to generate electricity for periods of higher demand. Furthermore, the transition process from high demand to low demand (or vice versa) cannot happen instantly, which results in a large amount of fuel waste in power stations and extra greenhouse gas emission. Even worse, the extra cost due to inefficient grid operation is distributed to consumers. With the advent of electric vehicles (EVs) and plug-in hybrid electric vehicles (PHEVs), the electricity requirement of the power grid will increase immensely [13, 14]. Peak and off-peak hours may have a more complicated and dynamic pattern in the future.

The solution provided in the smart grid is demand response, which can be applied to balance the supply and demand of electricity. Electricity usage during peak periods can be reduced or shifted in response to time-based rates and other forms of financial incentives. As a result, the power load is smoother in the smart grid, without sudden transitions. Thus the efficiency of the fossil-fuel power stations can be increased dramatically. Moreover, losses caused by energy theft, system failure, and other factors can be reduced in the smart grid because of the accurate and reliable automated monitoring systems [15, 16]. At first glance, such features enabled by the smart grid involve sacrifice on the part of consumers. Fortunately, consumers have opportunities to earn savings on their electricity bills in return. In addition to the pricing difference between peak and off-peak hours, the smart grid will allow utilities to launch more sophisticated programs as incentives. For example, some utilities may offer credits if consumers will allow a central office to control cycling their air conditioners on and off during times of peak power demand.

Consumers will use the grid in different ways. More consumers will become "prosumers"-both consumers and producers of energy [17]. Power will flow both ways, and other ancillary services may also be provided by these new prosumers. Some utilities even purchase back electricity generated by consumers. In theory, some consumers may receive checks from utilities, which could only happen in the smart grid. On top of potential savings, the smart grid offers consumers active control of their energy bills, allowing them to opt in and out of the demand response program; thus customer experience could be enhanced from just one-way communication.

1.1.3 Grid Reliability and Security Improvement

The current power grid has been improved quite a lot decade by decade in terms of reliability and security; however, blackouts still occur once or twice each year. The notorious blackout on Aug. 14, 2003 in parts of America and Canada affected 45 million people in the United States and 10 million people in Canada. The blackout was triggered by a relatively insignificant but overheated power line. In the smart grid, many intelligent sensors and actuators are deployed to monitor and control the grid's transmission system in real time [18–21]. New technologies in the smart grid, such as phasor measurement units (PMU), sample voltage and current many times per second, as opposed to once every few seconds in the current power grid. On systems equipped with smart grid communications technologies, system failure and hazardous situations can be reported to control centers promptly for fast reaction. As a result, distribution outages will be reduced in the smart grid. It would have been easier to detect the types of oscillations that led to the 2003 blackout in the smart grid. Moreover, advanced and comprehensive cybersecurity is provided in the smart grid communication infrastructures. Therefore, system reliability and security in the smart grid are greatly improved compared to the traditional power grid. However, the system will be even more complex.

With all of the new entities and energy resources, managing and optimizing the system will become increasingly challenging. Based on the aforementioned motivations, research and practical deployment in the smart grid need to achieve the objectives shown in the following list:

1) The smart grid needs to be adaptive to changing situations and able to self-heal when some system failures occur.
2) Customers are actively involved in the smart grid, based on dynamic pricing and other incentive programs.
3) The smart grid needs to increase the efficiency of grid operations while reducing losses.
4) The smart grid needs to be able to handle the integration of a large variety of generation options.

In order to achieve those objectives, it is necessary to develop and deploy communication infrastructures and advanced monitoring and control systems with cutting-edge technologies in the smart grid.

1.2 Smart Grid Communications Architecture

Its special publication *NIST Framework and Roadmap for Smart Grid Interoperability Standards, Release 3.0* [22], the National Institute of Standards and Technology (NIST) has defined that the smart grid is a complex cyber-physical system that must support 1) devices and systems developed independently by many different solution providers; 2) different utilities; 3) millions of industrial, business, and residential customers; and 4) different regulatory environments.

1.2.1   Conceptual Domain Model

A conceptual domain model was published in the NIST special publication to support planning, requirements development, documentation, and organization of the increasingly diverse collection of interconnected networks and equipment that will compose the smart grid, as illustrated in Figure 1.2. The NIST divides the smart grid into seven domains. Their roles and services in the smart grid conceptual model are described as follows:

1) **Customers**: the end users of electricity. May also generate, store, and manage the use of energy. Traditionally, three customer types are discussed, each with its own domain: residential, commercial, and industrial.
2) **Markets**: the operators of and participants in electricity markets.
3) **Service providers**: The organizations providing services to customers and to utilities.
4) **Operations**: the managers of the movement of electricity.
5) **Generation**: the generators of electricity. May also store energy for later distribution. This domain includes traditional sources (traditionally referred to as generation) and distributed energy resources (DER). At a logical level, "generation" includes coal, nuclear, and large-scale hydro generation systems that are usually

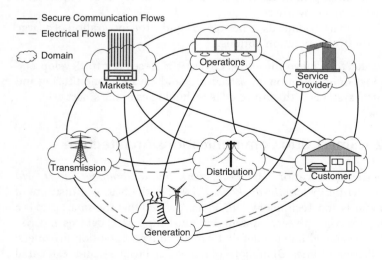

Figure 1.2 NIST conceptual domain model for the smart grid.

attached to transmission. Generation is associated with customer and distribution-domain-provided generation and storage and with service-provider-aggregated energy resources.

6) **Transmission**: the carriers of bulk electricity over long distances. May also store and generate electricity.

7) **Distribution**: the distributors of electricity to and from customers. May also store and generate electricity.

Each of the seven domains is a high-level grouping of physical entities that rely on or participate in similar types of services. In general, roles in the same domain have similar objectives. However, communications within the same domain may have different characteristics and may have to meet different requirements to achieve interoperability. The roles in a particular domain interact with roles in other domains to achieve interoperability.

1.2.2 Two-Way Communications Network

The interoperability that can be achieved in the NIST conceptual model for the smart grid is achieved by the secure communication flows that interconnect all seven domains. Generally speaking, it is a two-way communications network between utilities and customers. A high-level illustration of the smart grid communications network is shown in Figure 1.3. As it shows, the communications network in the smart grid consists of different types of networks and communication technologies that can be categorized into home-area networks (HAN), neighborhood-area networks (NAN), and wide-area networks (WAN).

- **Home-Area Networks in the Smart Grid**
  A HAN enables secure communication flows within a household. A gateway (usually a smart meter and a data aggregate unit) bridges a HAN to a NAN. Readers may have heard of the smart home that

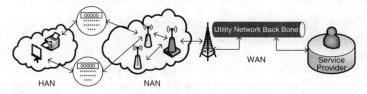

Figure 1.3 High-level illustration of the smart grid communication architecture.

can be achieved with intelligent and remote control. Note that a HAN in the smart grid communications network serves a different purpose than a smart home network. A smart grid HAN is part of the utilities' infrastructure that may or may not have Internet access. Even customers in the household may not have direct access to the network. The network that enables a smart home is owned by the customers and usually has Internet access; for example, a home Wi-Fi system. We are not ruling out the possibility that a smart grid HAN may merge with a smart home network in the future, especially when the Internet of Things and public Internet access will gain the support of utilities.

- **Neighborhood-Area Networks**
  A NAN is also referred to as the *last-mile* network to the customer side. It enables the secure flow of communication between households and the utilities' backbone network. Some existing NANs are composed of smart meters only, while others may utilize dedicated data aggregate units for relay. In either case, a smart meter is the gateway of a HAN that monitors and controls electricity consumption within that household. Through NANs and HANs, the smart grid achieves direct communications between customers and utilities. The communications are not only for metering and billing, but they also carry critical information for demand response, which actively controls the power load of some customers.

- **Wide-Area Networks**
  Utilities have their backbone network infrastructure that connects all major components, such as substations, power stations, and operation centers. This backbone network infrastructure is upgraded and reused as the WAN in the smart grid. The current power grid has applied power line carriers (PLC) for data communications for many years, especially in remote areas. Although the technology theoretically achieves wide area coverage, other communication technologies, such as fiber optics and cellular networks, are deployed to the smart grid to improve the WAN.

The three types of networks are named after their coverage. Nonetheless, their roles and services are vastly different; thus the communication technologies applied to each of the three types of network are based on their specific requirements. For example, a WAN is required to be high-speed, reliable, and secure while covering long-distance communications. Utilities have already deployed high-speed backhaul networks (also known as communication core networks) alongside most of their power transmission lines. The backhaul networks consist of optical fiber and Ethernet. PLCs [23, 24] are also applied in some areas to provide wide area communications. Most of PLC-based WANs are deployed for monitoring purposes instead of large data communications [23, 24]. Cellular communication technologies are also deployed as WANs for similar purposes. NANs and HANs are last-mile connections to customers [25, 26]. They are required to be low profile, low cost, reliable, and secure while providing enough bandwidth to meet latency requirements. In NANs and HANs, wireless communication technologies are preferred due to their flexibility and low-cost deployment. Wireless technologies for longer distance transmission, such as GPRS, WiMAX, and LTE, are promoted for communications between NANs and concentrators that connect to the backhaul network. Several wireless technologies for local-area networks, including Wi-Fi, Zigbee, and Bluetooth, are promoted for HANs and intracommuncations in NANs.

1.3 Applications and Requirements

In order to achieve the objectives of the smart grid, several important applications must be added to or upgraded in the current power grid. The most important applications and requirements include demand response, advanced metering infrastructure, wide-area situational awareness and wide-area monitoring systems, and communication networks and cybersecurity.

1.3.1 Demand Response

Demand response (DR) has been mentioned in earlier sections. It is the key component applied to the smart grid that can smooth the power load of the grid [27–30]. A smoother load may not necessarily reduce power consumption. It aims to decrease the gap between peak and off-peak grid loads and ease the transition process between high and low power demand, as shown in Figure 1.4. Maintaining a relatively steady power load would increase operational efficiency of renewable resources, as utilities worry only about a total demand that is given in advance. Fewer transition processes with less fluctuation would greatly reduce fossil-fuel waste and greenhouse gas emission from those types of power stations.

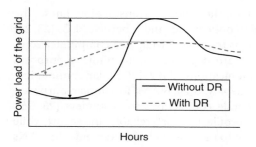

Figure 1.4 Smoother power load achieved by DR.

There are generally two ways to apply DR. One is direct load control from utilities, while the other is to actively involve customers by utilizing mechanisms and incentives. Both approaches have the goal of shifting some of the power demand from peak hours to off-peak hours. Direct load control can be implemented if consumers such as large business, government buildings, and some factories are willing to participate. Residential customers, on the other hand, may not be willing to give away their own control. Dynamic pricing motivates those types of customers in the smart grid. Generally speaking, electricity pricing is higher during peak hours. Customers adaptively control their appliances so that they can lower their bills. The incentive mechanisms are more complex than just a load shift in the smart grid. For example, in the past, the hours of 11 a.m. and 6 p.m. were set as off-peak hours in California with lower tariffs. However, the peak hours have shifted and extended over the years. Incentive mechanisms for DR in the smart grid need to accommodate these dynamic changes in a timely fashion.

1.3.2   Advanced Metering Infrastructure

Utilities and grid operators in the smart grid need to predict conditions in close to real time with sophisticated modeling and state estimation capabilities. Doing this will allow for more efficient dispatch and system balancing. This capability relies on advanced metering infrastructure (AMI) that carries near real-time data in the smart grid. Defined by the US DoE, AMI is an integrated system of smart meters, communications networks, and data management systems that enables two-way communication between utilities and customers, as shown in Figure 1.5. Customer systems include in-home displays, home-area networks, energy management systems,

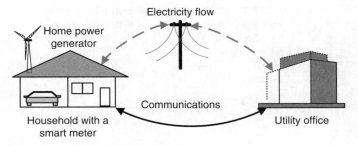

Figure 1.5 High-level illustration of AMI.

and other customer-side-of-the-meter equipment that enable smart grid functions in residential, commercial, and industrial facilities.

AMI is an ideal application of machine-to-machine (M2M) communications [31, 32] that achieves two-way communication between customers (through smart meters) and utilities [10, 33–36]. AMI equips each customer with a smart meter, whose basic function is to gather the energy consumption status and upload the information to the control center (also known as the power distributor or service provider). A smart meter is also capable of receiving control information (e.g. electricity pricing/tariff) from the control center. Such a two-way information exchange is near real-time. The information and control needed to implement residential demand response relies significantly on the development of AMI. AMI deployment in the smart grid will benefit both system operation and customer service.

1.3.3 Wide-Area Situational Awareness and Wide-Area Monitoring Systems

Wide-area situational awareness (WASA) and wide area monitoring systems (WAMS) in the smart grid monitor the status of power-system components over large geographic areas in near real time. The advanced monitoring system improves visibility and understanding of stresses in the power system and detects transient behavior that is not detected with traditional supervisory control and data acquisition (SCADA) systems.

As shown in Fig. 1.6, many types of intelligent electronic devices (IEDs) are deployed in the monitoring and control system, such as synchronized phasor measurement units (PMUs), phasor data concentrators (PDCs), circuit break monitors, solar flare detectors, etc. [37–42].

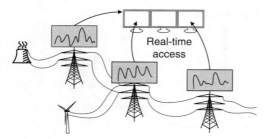

Figure 1.6 High-level illustration of the monitoring system.

Real-time access

Management of power-network components can be better achieved with the monitoring and control system. A system failure or a blackout can be ultimately anticipated, prevented, or quickly recovered from. While being considered a stand-alone system, the monitoring system is a composite of SCADA, AMI, energy management systems (EMS), and other systems in the smart grid that provides near real-time monitoring and control of the grid.

1.3.4 Communication Networks and Cybersecurity

The smart grid has a complicated and advanced communication infrastructure that may involve both private and public networks. In order to achieve interoperability between different domains, various types of communication technologies, including both wired and wireless technologies, are needed to support the infrastructure.

As mentioned earlier, the utility backbone network is mostly a private network deployed by utilities and grid operators, using fiber optics and Ethernet to meet the requirement of fast and reliable data delivery. Other parts of the communication infrastructure may not have fiber optics support, due to high cost of its development, implementation, and maintenance. For example, the "last mile" in AMI can be deployed with wireless technologies. A HAN may be supported by ZigBee, Wi-Fi, and other local area wireless technologies. A NAN may be supported by WiMAX, multihop Wi-Fi, etc. Communications in the smart grid monitoring system may be achieved by cellular networks or PLC. The exchange of information between the transmission and distribution systems will be automated and optimized by the development of standard data structures. Techniques from big data analytics and cloud computing will play a critical role in leveraging exponential growth in data.

With all those types of communication technologies and data involved, cybersecurity and control of access to the communication networks are critical issues to the smart grid [4, 43, 44]. Cybersecurity needs to be designed into the new systems that support the smart grid without impacting operations. In addition to the protection of information from traditional cyberattacks, smart grid cybersecurity must expand it focus to address the combination of information technology, industrial control systems, communication systems, and their integration with physical equipment and resources to maintain the security of the grid and to protect the privacy of consumers.

1.4 The Rest of the Book

In the rest of this book, we will explore some of the major topics in smart grid communication infrastructures and provide our solutions and suggestions. Some of the highlights are as follows:

- The overall communication infrastructures in the smart grid are studied.
- A complete information and communication technologies framework for the smart grid is proposed.
- The communication networks of the advanced metering infrastructure are studied.
  - A self-sustaining neighborhood-area network design is proposed.
  - An efficient power control scheme for the proposed network design is proposed.
- Demand response is studied, based on the communication infrastructures.
- Big data analytics and cloud computing are introduced into the smart grid communications to enhance grid operations and control.
- Network security is studied for smart grid communications.
  - Security schemes are proposed for communications in the advanced metering infrastructure.
  - ID-based security schemes are proposed for transmission over the Internet in the smart grid.

2

Smart Grid Communication Infrastructures

In this chapter, an information and communication technologies (ICT) framework will be explored to support the smart grid. The focus will be on the communication networks and their roles and requirements in the smart grid communication infrastructures.

2.1 An ICT Framework for the Smart Grid

2.1.1 Roles and Benefits of an ICT Framework

We have seen that the smart grid will have greatly improved communication networks compared to the traditional power grid. Two major achievements of the communication networks in the smart grid are 1) frequent and timely two-way communication capability between customers and utilities and 2) real-time monitoring and control of the vast majority of the power grid. In the current power grid, network communications are one-way only, with little information exchange. A better two-way communication network is required to control those detachable renewable energy sources, along with energy storage units, on the smart grid. Moreover, existing monitoring and controlling systems in the current power grid cannot provide the means to prevent system failures or blackouts such as the one in 2003. Given the massive scale and complexity of the smart grid, it is better to develop a unified ICT framework for the smart grid.

An ICT framework gives a clear view of the entire communication network and its integration with the physical components in the smart grid. It helps utilities to realize the interoperability of domains in the smart grid. An ICT framework is a step further from simple

Smart Grid Communication Infrastructures: Big Data, Cloud Computing, and Security,
First Edition. Feng Ye, Yi Qian, and Rose Qingyang Hu.

motivations of the smart grid. It lays a practical path for researchers and developers of the smart grid to follow for implementing features. For example, demand response (DR) is a promising feature in the smart grid that would improve the efficiency of grid operations by smoothing the power load. However, decisions and actions may not be made accurately in real time, even in the smart grid. Therefore, forecasting plays an important role in the smart grid. On one hand, an energy forecast helps power generators to plan electricity generation ahead of time. Thus fuel waste due to sudden transitions between different loads can be reduced. On the other hand, a pricing forecast helps customers to schedule their electricity usage more economically [45, 46]. How to model and achieve pricing forecast and energy forecast is the practical problem to be addressed. An ICT framework would reveal the information flow between the domains of customers, utilities, and power stations.

An ICT framework would also provide required technologies and equipment to implement features such as DR in the smart grid. Several existing research works on DR rely on real-time power consumption information from metering data and real-time response from both power generators and customers. Some research works assume precise power requests from customers instead. However, all of the assumptions are hard to achieve in practice. Moreover, with renewable energy sources that are difficult to control and detachable microgrids [47–49], an optimal control of the power grid is extremely hard and expensive to implementwith just the information and computing resources from the utilities. An ICT framework would reveal such issues and allow grid planners to add extra tools, such as cloud computing and big data anlytics.

2.1.2 An Overview of the Proposed ICT Framework

Figure 2.1 shows an overview of the proposed ICT framework. The ultimate purpose of the proposed ICT framework is to develop systems that can improve the efficiency and reliability of the smart grid. To make the illustration clearer, our proposed ICT framework is described as developing DR with incentive mechanisms that are based on dynamic pricing. Entities, components, and their roles are intended to enable *energy forecasts* for power generators, and *pricing/tariff forecasts* (pricing forecast hereafter) for customers. It is intuitive to research and develop other systems, such as real-time monitoring

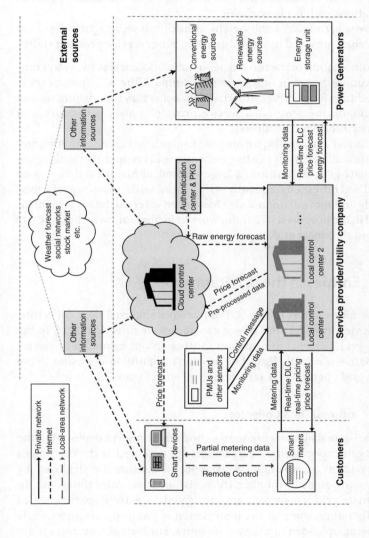

Figure 2.1 An overview of the proposed ICT framework.

and controlling systems, within the proposed ICT framework. Three types of networks are applied in this framework:

- *Local-area networks*, established for customers to enable communications within a household;
- *Private networks*, established by utilities and service providers; and
- *Internet*, provided by a third-party Internet service provider (ISP).

The combination of the three types of networks enables two-way communications between utilities and customers. The ICT framework is divided into four entities: internal data collectors (i.e. customers and grid monitoring sensors), a service provider, power generators, and external information sources.

In the ICT framework, we propose to combine cloud computing and big data analytics with existing networks and computing resources in the smart grid. As a result, a large amount of historical data can be stored and refreshed frequently, so big data analytics can be performed quickly and more economically. Moreover, external data is also introduced in the proposed ICT framework to further improve the forecasts made in the smart grid.

2.2 Entities in the ICT Framework

Of the four entities in the ICT framework, internal data collectors, the service provider, and power generators are directly related to the smart grid. External information sources do not belong to the smart grid natively; nonetheless, they provide insightful information to the smart grid operations. Examples are given in Figure 2.2.

2.2.1 Internal Data Collectors

Internal data collectors are sensors and smart meters deployed in the smart grid. Specifically, smart meters are deployed at the customer's site by utility companies. Many types of sensors are deployed by utility companies to monitor transmission lines, substations, etc. In the smart grid, customers are motivated to actively participate in DR. Therefore, some of the information is given to customers while also being uploaded to utility companies. For instance, customers are granted access to the electricity usage data of their own properties. In many cases, customers themselves have established local-area

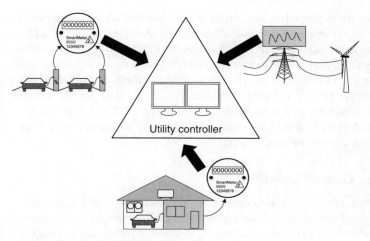

Figure 2.2 Examples of internal data collectors.

networks (LANs) or wireless LANs (WLANs), based on Wi-Fi, Bluetooth, etc., that connect smart devices, smart appliances, and corresponding smart meters in each household. Smart devices can be smart phones, tablets, laptops, etc. Without loss of generality, smart phones will be used to indicate smart devices. From smart phones, customers are able to monitor electricity consumption of their appliances and possibly control them. In practice, LANs/WLANs established by customers usually have access to the Internet. Consequently, remote monitoring and remote control can be applied by customers anywhere they have Internet access.

Apart from fixed household appliances, electric vehicles (EVs) and plug-in hybrid electric vehicles (PHEVs) are mobile appliances that have more resilient electricity requirements. For example, an EV can be charged in a household, but it can also be charged in a public charging station or a capable parking lot. Furthermore, some companies and researchers are pushing to standardize batteries for EVs. In that case, customers can go to a battery exchange station and replace depleted battery with a fully charged one. The electricity consumption of EVs at charging ports will be captured by corresponding smart meters or specific types of sensors. Other useful information, such as the location and possible routes of EVs, may be gathered by some external agents with the customers' permission. Such data will be considered as external data by the ICT framework.

Sensors deployed in the smart grid transmission line monitoring systems are internal data collectors as well. A typical sensor is a phasor measurement unit (PMU) that is used in wide-area measurement systems (WAMS) to measure the electric waves on an electricity grid using a common time source for synchronization. Phasor data concentrators (PDC) are deployed to collect PMU data for the central control facility, and control and monitor the PMUs. For 60 Hz systems, PMUs must deliver between 10 and 30 synchronous reports per second, depending on the application [50].

2.2.2 Control Centers

Control centers are deployed and operated by utility companies. Specifically, there are three types of control centers in the proposed framework: local control centers, a cloud control center, and an authentication server (AS) with a private-key generator (PKG). For simplicity, AS or PKG will be used interchangeably to indicate the group hereafter. It is also reasonable to assume that the PKG is a trusted third party. In terms of responsibility, control centers as a whole unit make the *energy forecast* to power generators, and the *pricing forecast* to customers. In DR, a control center is also responsible for direct load control (DLC) of both power consumption by customers and electricity generation by power generators. DLC usually applies to power companies themselves. For instance, they cycle air conditioners and water heaters on and off during periods of peak demand to smooth the power generation load.

The smart grid has a massive scale in almost every way: number of users, area of coverage, complexity of its infrastructure, etc. Therefore, local control centers need to be distributed across the power grid for better scalability and reliability in the proposed ICT framework. For instance, a local control center can be deployed close to or inside a power distribution substation. While each substation covers a relatively small area, a local control center is responsible for the customers within that particular area. Substations are currently connected by a high-speed private backhaul network deployed by the utility company. With extra gateways to the backhaul network, local control centers are interconnected through the utilities' backbone networks. Although local control centers and private networks are controlled by utility companies, it is better to isolate data storage so that local control centers do not share collected data during normal operations. Some necessary conditions for a local control center to share or pass data to other local control centers include routine maintenance or temporary system downtime due to cyberattacks or natural disasters. The main functions of local control centers include:

1) internal data (i.e. metering data and sensor data) collection;
2) preprocessing sensitive data to protect the privacy of customers;
3) real-time direct control of the power grid;
4) finalizing energy forecasts for power generators;
5) generating price forecasts for customers.

Unlike local control centers, the cloud control center is a comprehensive unit composed of complicated and distributed hardware as well as software. Different types of services, such as infrastructure as a service, platform as a service, and software as a service, can be provided. Nonetheless, such complexity needs to be transparent to users (i.e. the utility company). Therefore, the cloud control center can be viewed as a powerful single unit in the framework. As shown in Figure 2.3, the cloud service provider serves the central power grid, detachable microgrids, customers' smart devices, and many other types of customers that are not part of the smart grid. Deploying and maintaining a large cloud computing center could be too expensive for

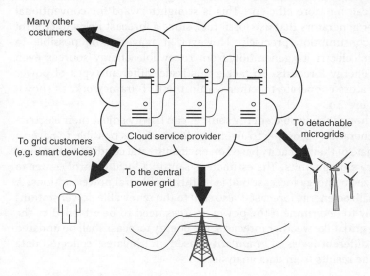

Figure 2.3 Cloud computing service and the power grid.

utility companies. Thus it is reasonable to rent a cloud control center from a public cloud service provider at the cost of losing control of the data connection links, because the cloud control center is connected internally by high-speed networks that are not controlled by the utility company. The main functions of the cloud control center in the ICT framework include:

1) store data uploaded from local control center for a certain period;
2) fetch data from external sources;
3) perform big data analytics on collected data and make *raw energy forecasts* for each area and the entire grid.

The raw energy forecast is sent back to local control centers for finalizing. Utilities may not fully trust the public cloud services and thus retain key features for finalizing energy forecast.

2.2.3 Power Generators

Power generators consist of central conventional energy sources and detachable renewable energy sources. If an energy forecast is provided for a sufficient time period to the grid operators, the transition between peak and off-peak electricity generation is smoother and thus can be more efficient. This is straightforward for conventional power generators that use fossil fuel, since a forecast helps to control fuel consumption precisely. However, it may not be possible to control electricity generation from renewable energy sources even with energy forecasts. To tackle this issue, different types of power generators cooperate together within the ICT framework, as shown in Figure 2.4.

Although renewable energy sources can hardly adjust their electricity generation according to an energy forecast, it is possible to have an estimate of their capacity based on environmental conditions and their energy storage units. The estimate is sent to a local control center to finalize an energy forecast that is mainly for central power stations. A specialized energy forecast is also sent to the renewable power stations mainly to determine if the power sources need to be attached to the main grid. Like weather forecasts, the energy forecast shall be updated with different levels of granularity based on the latest collected data and the results from data analysis.

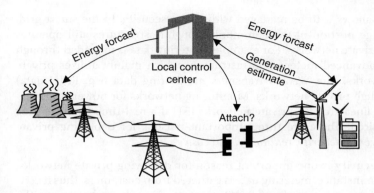

Figure 2.4 Example of power generators in the ICT framework.

2.2.4 External Data Sources

Due to many uncertainties, such as the generation capacity of renewable energy sources and consumer power usage patterns, internal data from the grid alone is not enough to make accurate energy and pricing forecasts. In order to improve research and development of smart grid features, different types of data from external sources are included in the ICT framework. For example, weather forecasts can provide better estimates of amount of electricity generated by renewable energy sources. Locations and routes of EVs can be used to estimate their energy consumption as well as their schedule. Useful information can also be extracted from many other external sources. The smart grid will certainly operate more efficiently with all this external data.

2.3 Communication Networks and Technologies

2.3.1 Private and Public Networks

Communications in the smart grid are achieved through *private* and *public* networks. Private networks are deployed and maintained by utilities. Public networks are provided by third-party service providers, such as cellular connections and the Internet. In the current power grid, almost the entire network is private, because that allows it to be managed with better security. In the smart grid, a large portion of the communication infrastructures still operates on private networks. For example, metering data is uploaded through the advanced metering infrastructure which mainly utilizes private networks. The WAMS transmits monitoring data (e.g. PMU data) through private networks. Monitoring networks for power transmission lines are also private networks [51] if a real-time connection is required. There are three important reasons for deploying private networks: *security, reliability*, and *cost*.

- **Security** is one important reason for deploying private networks. For instance, metering data is gathered from customers, thus it contains much private and sensitive information about customers. The lifestyle of a customer may be revealed if metering data is leaked.
- **Reliability** is another important issue. For instance, real-time monitoring in the smart grid requires low latency (e.g. 10 to 100 ms for PMUs in WAMS). Public cellular networks or Internet service providers can hardly meet the latency requirement, due to their complicated protocols and mechanisms. The private networks in the smart grid are designed specifically to fulfill the latency requirement.
- **Cost** is also a reason for private networks. Subscription fees for public network service providers can be overwhelmingly high in the long run. Moreover, the power grid may cover a larger area than a public network service does. For instance, power transmission line monitoring networks may cover remote areas that have no public network access. The private networks are built on different types of communication technologies, including various types of wireless networks (e.g. Wi-Fi, Zigbee, WiMAX, etc.) and high-speed wired networks (e.g. fiber optics and Ethernet) [52]. Specific technologies are chosen to balance optimal network performance with hardware and maintenance costs.

2.3.2 Communication Technologies

Without loss of generality, we will discuss communication technologies applied to AMI and WAMS for illustration. AMI and WAMS constitute most of the communication infrastructure in the smart grid.

Other communication networks can be explored based on the ones applied to AMI and WAMS with slight modifications.

AMI is one of the most important infrastructures in the smart grid. The communication networks of AMI in the smart grid generally consist of home-area networks (HANs), neighborhood-area networks (NANs), and a wide-area network (WAN). An overview of AMI is shown in Figure 2.5. A HAN is established within a household, connecting a smart meter and smart appliances with sensors and actuators. Smart meters upload metering data to data aggregate units (DAPs) that are deployed in a neighborhood. A NAN is formed by DAPs and smart meters in a neighborhood. Metering data is finally gathered by the metering data management system (MDMS) through the high-speed WAN. MDMS will provide storage, management, and processing of meter data for proper usage by other power system applications and services.

Wireless technologies have been widely used in the last-mile communication systems (i.e. HANs and NANs) in the smart grid [25, 26, 53–55] because of their flexibility in deployment and environmental adaptiveness, especially in extreme situations. A HAN is similar to a local-area network. Numerous works have been proposed

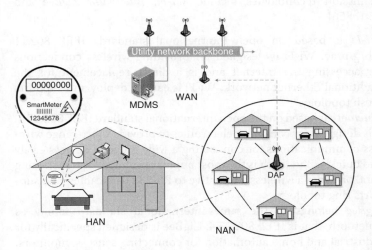

Figure 2.5 High-level illustration of the advanced metering infrastructure.

Table 2.1 Enabling wireless technologies in HANs.

Key Criteria	Wi-Fi	Bluetooth	ZigBee	Z-Wave
Frequency band	2.4 GHz	2.4 GHz	868/915 MHz	900 MHz
	5 GHz		2.4 GHz	
Speed	54 Mbps (b/g)	2.1 Mbps (V2.0)	250 Kbps	9.6 Kbps
	150 Mbps (n)	25 Mbps (V3.0/4.0)		
	433 Mbps (ac)	50 Mbps (V5.0)		
Range	70 m (in)	10 m	70 m (in)	30 m (in)
	250 m (out)		450 m (out)	100 m (out)
Power	High	Lower	Lowest	Lowest
Maximum nodes	2007	8	over 64,000	232

to study HAN as a network with a small coverage area and a low rate of data transmission, including a single smart meter as the collector and several smart appliances. Both wireless and wire-line technologies can be applied to HANs. As shown in Table 2.1, wireless transmission technologies include *Wi-Fi, Bluetooth, ZigBee* and *Z-Wave* [56].

- *Wi-Fi* is based on open international standards IEEE 802.11 a/b/g/n/ac. Wi-Fi is designed for providing wireless connections for accessing the Internet and is a direct replacement for the traditional Ethernet network. Wi-Fi is easy to deploy and supports mesh topology.
- *Bluetooth* is based on the open international standard IEEE 802.15.1. It is designed for consumer electronics to provide short-range wireless communications for connecting a wide range of devices easily and quickly. Bluetooth is the most popular protocol for transferring data, provides a wireless alternative to RS-232 data cables, and supports ring topology.
- *ZigBee* belongs to the same international standard family as Bluetooth, that is IEEE 802.15.4. ZigBee is designed specifically for industrial and home automation for connecting sensors, monitors, and control devices. ZigBee has low power requirements and implementation costs. It also supports mesh topology.

- *Z-Wave* is a proprietary wireless standard designed for home control automation, specifically for remote control applications in residences. It was originally developed by Zensys A/S and is being marketed by Z-Wave Alliance. Z-Wave wireless protocol provides reliable and low-latency communication of small data packets within HANs. Z-Wave has low power, low latency, and low cost. It has less interference due to its use of sub-GHz frequency. It also has higher propagation range (i.e. is 2.5 times that of a 2.4 GHz signal) and supports mesh topology.

Short-range wired communication technologies are also candidates for HANs. As shown in Table 2.2, some candidates are *X10*, *HomePlug GP*, *BACnet* and *KNX*.

- *X10* is an open standard that is designed for providing simple automation functionality, such as on and off.
- *HomePlug GP* is based on international standard IEEE 1901. It is designed for consumer electronics to provide short-range wireless communication to connect a wide range of devices easily and quickly.
- *BACnet* is based on the open standard ISO 16484-5 and ANSI/ASHRAE 135-2008. It is a data communication protocol that attempts to unifies all the proprietary communication protocols into single communication language.
- *KNX* is based on standards IOS/IEC 14543-3, GB/Z 20965, ANSI/ASHRAE 135, CENELEC EN 50090 and CEN EN 13321-1. It is a global standard protocol designed basically for home automation and control.

Table 2.2 Enabling wired technologies in HANs.

Key Criteria	X10	HomePlug GP	BACnet	KNX
Wireless Support	Yes	Yes	No	Yes
	310 MHz	ZigBee		KNX RF
	433 MHz			868.3 MHz
Speed	20 bps	4–10 Mbps	Depends on LAN used	9.6 kbps (wired)
				16.4 kbps (wireless)
Maximum nodes	256	253 (Theoretically) 10 (Practically)	No limit	57,600 (wired)

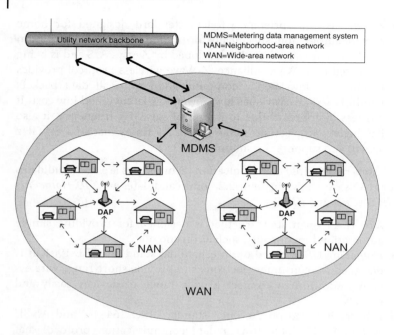

Figure 2.6 High-level illustration of NANs.

A NAN is a bridge connecting HANs and the WAN, as shown in Figure 2.6. It consists of many DAPs, each collects data from several smart meters and transmits the data to a concentrator through NAN gateways in their uplink transmissions. The DAPs also broadcast the control information from the MDMS to their surrounding smart meters. Specific control messages to a certain smart meter are also transmitted through DAPs. A DAP communicates with its surrounding smart meters with the same technologies that form a HAN. Internal communications in a NAN (i.e. communications between DAPs) are proposed to be accomplished by Wi-Fi (especially by following *IEEE* 802.11s) because a NAN is a wireless mesh network with high data transmission rates. Some of the DAPs are also equipped with cellular technologies such as WiMAX or LTE for reliable longer-distance transmission to the concentrator (i.e. a gateway to backhaul networks). Besides wireless technologies, wired technologies are also applied to NANs. A widely used wired technology is PLC, which may operate in ultra-narrow band, narrow band, and broadband. Frequency band information is given in Table 2.3.

Table 2.3 PLC operating frequency bands.

Type	Frequency bands
Ultra narrow band PLC	0.3 *kHz*–3 *kHz* bands
Narrow band PLC	0.3 *kHz*–3 *kHz* bands
Broadband PLC	1.8 *MHz*–250 *MHz* bands

The WANs are part of the backhaul networks in the smart grid. PLC, Internet protocol (IP)-based networks, and some wireless technologies can be applied to WANs. The pros and cons of each technology are given below:

- **Pros of PLC in WANs**
  - Provides a low-cost solution to overlay the communication network over already available power lines.
  - Complete control over the communication path with extensive coverage that is solely controlled by the utility industry.
  - Provides a direct route between controllers and other subsystem to ensure low latency.
  - Mature technology with many variants available commercially.
- **Cons of PLC in WANs**
  - Power lines are connected to various kinds of equipment, such as motors or power supplies, that can act as noise sources that eventually degrade the performance of the PLC.
  - The load impedance fluctuation and electromagnetic interference cause signal attenuation and distortion, which can result in the failure of the communication link.
  - Lack of standards and government regulation due to industry fragmentation result in high interference from other PLC technologies deployed at close range.
  - Costs of PLC modems are still high.
  - There is a coexistence issue with many commercial technologies.
- **Pros of IP-based networks in WANs**
  - They have rich convergence capabilities that can help to connect the overall systems and subsystems in smart grid
  - They can provide quality of service and reliable connection using technologies such as DiffServ and MPLS.
  - Security can be enhanced using known technologies (i.e. IPSec).

- **Cons of IP-based networks in WANs:**
  - In the case of a master/slave configuration, transmitting IP packets from the slave is not possible, which might increase the data latency for those applications that require fast response, as smart grid does.
  - Unless a private IP-based network (e.g. Internet VPN) is used, security remains a crucial issue.
- **Pros of wireless networks in WANs**
  - They provide a huge coverage area, potentially at a low cost
  - Cellular data transmission has lower costs and much higher data rates.
  - Some wireless technologies (e.g. WiMAX) can support mesh networks for higher reliability.
- **Cons of wireless networks in WANs**
  - Utilities have to depend on these technologies without any control over them.
  - A connection to the network is required before data can be transmitted, which may be a problem in case of outages and emergencies.
  - Longer latency degrades real-time services.

Besides private networks, smart grid communications also utilize the Internet. PMUs and PDCs in WAMS can use ubiquitous cellular networks, such as GPRS, UMTS, and LTE, for data transfers. Smart appliances may connect to an established Wi-Fi network with Internet access. Public cloud computing service can be involved in the smart grid, and use of the Internet is inevitable for communications from and to cloud computing services. However, public cellular networks or ISPs can hardly meet the latency requirements due to complicated protocols and mechanisms, which limits their use in real-time protective systems.

2.4 Data Communication Requirements

Communications in the smart grid have various requirements. Some of the major requirements are *latency and bandwidth, interoperability, scalability*, and *security*. Those requirements must be considered together in the design and deployment of smart grid communications along with other issues, such as reliability, availability, and cost.

2.4.1 Latency and Bandwidth

Several types of data are transmitted in the smart grid with different latency requirements [57]. One characteristics of communication is that many interactions must take place in real time, with hard time bounds. The real-time operational data communications in the smart grid include online sensor/meter reading and power system control signals. Power system control signals mainly include supervisory control of the power process on secondary or higher levels. Several applications and their corresponding latency requirements in the smart grid are listed in Table 2.4.

As more and more interconnected intelligent elements are added to the smart grid, the communication infrastructure should be able to transport more and more data simultaneously without severe impacts on latency. Therefore, the network bandwidth must increase faster than the demand of these interconnected intelligent elements in the network. For example, a distribution substation connected to 10,000 feeders with each feeder attached to 10 customers could require a bandwidth that is over 100 Mbps if a message is generated

Table 2.4 Data communication latency requirements.

Application	Latency Requirement
Demand response	2 s–5 min
Remote connect/disconnect	2–5 s
AMI, real-time pricing	2 s
Metering data management	
Outage management	300 ms–2 s
In-home displays	
Emergency response	10–100 ms
Synchro-phasor	10–100 ms
DA protection event notification	3–10 ms

every second. With more frequent message transmissions, bandwidth should be expected to increase in smart grid communications.

2.4.2 Interoperability

Interoperability is the ability of two or more networks, systems, devices, applications, or components to communicate and operate together effectively and securely, without significant user intervention. To achieve interoperability, smart grid communications require agreement on a physical interface and communication protocols. In addition, exchanging meaningful and actionable information requires common definitions of terms and agreed-upon responses. Moreover, consistent performance requires standards for the reliability, integrity, and security of communications.

2.4.3 Scalability

Scalability is another critical issue in smart grid communications. The smart grid communications system involves huge numbers of customers and countless sensors, and it also covers vast geographical areas. For example, the MDMS is considered a centralized control center in AMI. However, in practice, it is better to deploy it in a distributed way to achieve efficient communications that can meet the latency requirements. The amount of data to be gathered and processed in the AMI is massive, and customers are spread across cities and the countryside.

2.4.4 Security

Finally, network security is inevitable in smart grid communications. Various pieces of data in smart grid communications may contain much private information about customers, as well as control messages that operate the grid. Leakage of metering data will jeopardize the privacy of customers. Forging or manipulating control data may hamper optimal demand response. In addition, alteration of sensing data from WAMS may lead to severe system failures or blackouts.

2.5   Summary

In this chapter, we presented an information and communication technologies framework for the smart grid. The ICT framework is intended to show the pieces of the entire communication network and their integration with physical components in the smart grid so that utilities can improve interoperability of different domains. In addition, requirements for possible technologies of communication networks to support the smart grid were also discussed in this chapter. It is recommended that readers get a better understanding of the specific communication technologies mentioned in this chapter, such as Wi-Fi, Bluetooth, ZigBee, WiMAX, PLC, etc.

3

Self-Sustaining Wireless Neighborhood-Area Network Design

Wireless technology has been widely considered for last-mile communications in the smart grid, because of its growing performance and the relatively low cost but more flexible nature. In the advanced metering infrastructure (AMI), both home-area networks (HANs) and neighborhood-area networks (NANs) can be deployed using wireless technologies. While HAN has been widely studied using either IEEE 802.11 (Wi-Fi) or IEEE 802.15.4 (ZigBee) due to its small coverage and low data rate transmission [53, 54], the standalone NAN should be explored. In this chapter, we propose a self-sustaining wireless NAN design for AMI in the smart grid. We then propose an optimization approach to achieve minimum total cost for the design. To optimize the system, we first study the optimal number of gateway DAPs. Then we propose two geographical deployment methods in order to achieve fairness for customers. To further enhance fairness, we set different transmission power levels for gateway DAPs. We also achieve the global uplink transmission power efficiency. Compared with the existing game theoretical approach, our approach can increase the energy efficiency of the system. To quickly approach the optimal result, we propose an algorithm for which the computational complexity depends solely on solving a linear system.

3.1 Overview of the Proposed NAN

3.1.1 Background and Motivation of a Self-Sustaining Wireless NAN

In a neighborhood powered by the smart grid, each household or apartment (we will use only household hereafter since the DAPs are

Smart Grid Communication Infrastructures: Big Data, Cloud Computing, and Security,
First Edition. Feng Ye, Yi Qian, and Rose Qingyang Hu.
© 2018 John Wiley & Sons Ltd. Published 2018 by John Wiley & Sons Ltd.

not affected by the types of customers) is equipped with a smart meter. Besides residential customers, commercial and industrial customers are also equipped with smart meters in their buildings and parking lots with electric vehicle charging ports. Each smart meter collects facility status and customer-side power usage. Smart meters are connected to the service provider through the NAN and WAN to upload customer information to the service provider. The service provider sends real time pricing and tariff information to all the smart meters through downlinks.

A NAN consists of multiple DAPs. Each DAP collects data from a group of nearby smart meters. Unlike a HAN, which is deployed in a protected household, a NAN is most likely to be deployed in an open public area; it must be low-profile and flexible so it can be easily deployed, moved, or replaced. Unlike a WAN, which is powered by the grid directly and has a wired connection to satisfy its Gbps-level data rate, a wireless-based NAN has a much lower data transmission rate. Fortunately, a NAN does not require a Gbps-level data rate. In practice, a smart meter generates about 12 KBytes of data during one sampling period [34]. In that case, short-range wireless technology such as Wi-Fi would provide sufficient performance for local transmission within a NAN [58]. Moreover, operating a wireless-based NAN does not require too much power, which makes it possible to apply renewable energy (e.g. solar power) to the NAN design.

However, Wi-Fi is not a good solution to longer distances (e.g. uplink/downlink transmission of NAN gateways to/from the concentrator). A solution is to use cellular communication technologies, for instance IEEE 802.16 (WiMAX) [59], for NAN gateways. Note that the term *gateway* is used interchangeably with *NAN gateway* in the rest of the chapter. Ideally, WiMAX provides a data rate up to 70 Mbps and distance up to 48 kilometers [60]. Although those two cannot be achieved at the same time, a high data rate can be achieved when the distance is relatively short in a neighborhood. WiMAX also has an advantage over other types of longer distance wireless transmission (e.g. LTE) in that it can be operated in an unlicensed spectrum under FCC rules [61].

Since a larger number of hops in a mesh network results in a longer delay [62], some gateways need to be deployed in the NAN. In addition to reducing end-to-end delay, multiple gateways would also enhance the reliability of the NAN [63]. However, a larger number of gateways does not necessarily produce better transmission performance. It is important to find the optimal number of gateways in a NAN. In this design, we will focus on the scenario with a single concentrator.

3.1.2 Structure of the Proposed NAN

The proposed self-sustaining wireless NAN design is based on an IEEE 802.11s-enabled (or Wi-Fi based) [64] wireless mesh structure. An overview of the NAN design is shown in Figure 3.1. The NAN consists of a concentrator and a number of DAPs. The concentrator is deployed close to the neighborhood. It is connected to the wired backhaul network for communicating with the service provider. DAPs are deployed to cover the entire neighborhood. Each of them collects data from several surrounding smart meters and uploads the data to the concentrator. Each DAP also broadcasts controlling information from the service provider to those smart meters.

IEEE 802.15.4 (ZigBee) has been widely adopted for HANs and thus is assumed to be the technology used for communications within a HAN and between a HAN and a DAP [64]. Unlike the single-hop low data rate transmissions between a HAN and a DAP, local NAN transmissions among DAPs introduce multihop wireless mesh networking. Therefore it requires a higher data rate for the transmissions. IEEE 802.11s (Wi-Fi-based wireless mesh network) is

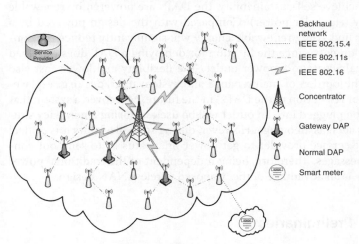

Figure 3.1 Overview of the proposed NAN structure.

a good candidate, since it meets these criteria. Among all the DAPs, some of them are chosen as the gateway DAPs which communicate with the concentrator directly. Although Wi-Fi performs well in local NAN transmission, its transmission range is limited; therefore it can hardly support uplink and downlink transmission between gateway DAPs and the concentrator at longer distances (i.e. hundreds of meters). As a solution for gateway DAPs, we adopt IEEE 802.16 (WiMAX) as the technology for transmission. Using WiMAX has an extra benefit compared with similar technologies (e.g. UMTS or LTE). WiMAX can be deployed using an unlicensed band (e.g. 5.8 *GHz*) in order to lower the service cost by avoiding license fees for band access. However, the use of unlicensed bandwidth must follow certain restrictions imposed by the FCC [61].

Normal DAPs are deployed in the neighborhood to cover all customers. The number and the geographical positions of gateway DAPs must be carefully chosen so that the data from all the customers can reach the concentrator in time. Traditionally, a network may only have one or multiple gateways that are located closer to the destination compared with other units within this network. However, multiple gateways need to be deployed sparsely in a NAN because of its wide-ranging latency requirement (3 ms to 5 min) [65, 66]. If some metering data has too many hops to reach a gateway, it may not get delivered within the latency requirement.

To achieve self-sustainability, the DAPs are powered by renewable energy (e.g. solar power). Compared with the design powered by a power line, using renewable energy will significantly reduce the construction cost, since the location for deploying a DAP does not need physical access to a power outlet. The flexibility and lower cost also ease the burden of maintenance and system upgrades (e.g. replacement or relocation of the DAPs) in the future. Moreover, a system that must be plugged into an outlet will be useless during emergency situations such as an outage. However, during such blackouts it is critical for the service provider to get power line status or to send out control messages. Therefore, being independent of the traditional power supply is an advantage of the proposed wireless NAN design.

3.2 Preliminaries

In this section, some preliminaries are presented so that design details of the proposed NAN can be better described. We assume each DAP is powered by solar power and a backup battery with limited capacity. Thus we will show the modeling of solar panel charging and battery life cycle. We will also mention the path loss model used for wireless transmissions in the NAN design.

3.2.1 Charging Rate Estimate

The charging rate estimate is modeled for a solar panel that powers a DAP. For simplicity, we assume that the maximum charging rate occurs at noon (assuming 12:00 p.m.) and the minimum charging rate occurs at sunrise and sunset. Moreover, let the charging rate be symmetric during the daytime. Then the charging rate at time t with perfect weather conditions is estimated as follows:

$$c(t) = \frac{w}{\sqrt{2\pi\sigma^2}} e^{-\frac{(t-\mu)^2}{2\sigma^2}}, \ t_r \leq t \leq t_s \tag{3.1}$$

where $\mu = 13$, σ is the parameter that determines the max / min (e.g. noon/sunrise) charging ratio, w is the weight factor depending on number of the solar panels and weather conditions, and t_r, t_s are sunrise time and sunset time respectively. In other words, σ is a value determined by the nature of the solar panel, and w is determined by the size of solar panel.

For example, as stated in [67], the max / min charging ratio with sunny weather conditions is about 2.94. Even with cloudy weather conditions, the solar panel can still generate electricity with a max / min charging ratio of about 2.33. With these facts, we find $\sigma_s = 3.40$ and $\sigma_c = 3.84$ for sunny and cloudy weather conditions respectively. Although a solar panel still functions with imperfect weather conditions, it generates electricity at a significantly lower rate. The ratio of the maximum charging rate during sunny weather to that of cloudy weather is about 6.71. Equivalently speaking, we have $w_s/\sigma_s = 6.71 w_c/\sigma_c$. If a unit solar panel has a maximum charging rate of 1 W during sunny weather, then its $w_s = 8.53$ and $w_c = 1.44$. The charging characteristics during perfect weather conditions of such a solar panel unit are shown in Figure 3.2. Without loss of generality, *cloudy weather* is applied to illustrate all bad weather, such as rainy or foggy conditions.

Although the performance of a solar panel varies in different seasons, we assume the size needed is determined based on the worst season (normally winter). In the rest of the paper, for simplicity we assume the sunny weather and cloudy weather conditions are those of typical days in winter.

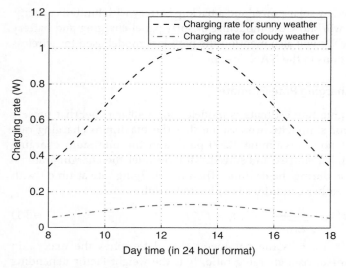

Figure 3.2 Solar panel charging rate estimate.

3.2.2 Battery-Related Issues

The first issue for a battery is the charging cycle. The battery has to be replaced if its state of health (SoH) falls below $SoH_{dead} = 80\%$ of rated capacity [68]. The SoH is updated after one charge; for simplicity, we adopt the linear estimation as:

$$SoH_{new} = SoH_{prev} - (1 - SoH_{dead}) \times \frac{1}{Cy_{max}} \times \frac{C_R}{C_{eff}}, \qquad (3.2)$$

where Cy_{max} is the maximum battery life cycle, C_R is the rated capacity, and C_{eff} is the effective capacity highly effected by discharging current $I_{discharge}$ for discharging time T (typically 20 hours), where

$$C_{eff} = C_R \times \left(\frac{C_R}{I_{discharge} \times T} \right)^{0.15} \times SoH\%. \qquad (3.3)$$

However, the discharging current for a DAP is relatively small (< 1 A), and thus $C_{eff} \approx C_R$. In the following discussion, the effective capacity is thus omitted. To make the illustration clearer, let the deterioration of health ($DoH = 1 - SoH$) be the status of the battery to monitor. And thus $DoH_{dead} = 1 - SoH_{dead}$ accordingly. We rewrite Eq. (3.2) as:

$$DoH_{new} = DoH_{prev} + (DoH_{dead}) \times \frac{1}{Cy_{max}}. \tag{3.4}$$

For simplicity, we assume that the maximum battery life cycles Cy_{max} depends solely on the depth of discharge (DoD); for example, when $DoD = 80\%$, $Cy_{max} = 500$.

Table 3.1 Selected DoDs and their maximum cycles.

Depth of discharge (%)	Maximum cycles
100	300
80	500
50	1200
25	2000
10	4000

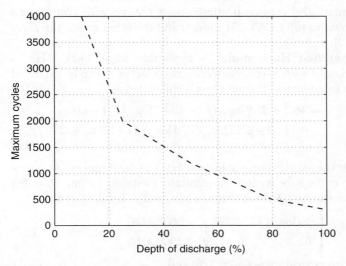

Figure 3.3 Modeling of maximum cycles against depth of discharge.

$$Cy_{max} = \begin{cases} -150DoD + 5500 & 10 \le DoD < 20 \\ -43.33DoD + 3366.67 & 20 \le DoD < 50 \\ -23.33DoD + 2366.67 & 50 \le DoD < 80 \\ -10DoD + 1300 & 80 \le DoD < 100 \end{cases} \quad (3.5)$$

We model the maximum cycles under different DoD with several linear segments based on data given in Table 3.1 [69]. The estimates are calculated by Eq. (3.5). Figure 3.3 also illustrates the estimates of maximum cycles with different depths of discharge.

3.2.3 Path Loss Model

The uplink transmission from a gateway DAP to the concentrator is usually further than a close-in reference distance (e.g. 100 m), and it is much further than a single-hop local transmission; therefore we need to apply path loss to the transmission model. Path loss is not considered in Wi-Fi transmission within the NAN. Wi-Fi transmission loss has been considered as the receiving data rate being significantly lower than theoretical limit of the physical layer data rate. There are two well-known path loss models used to estimate the WiMAX path loss in smart grid communication: one is *the modified Hata model* (also known as the COST 231 model) and the other one is the *Erceg model* [70].

The **modified Hata model** is applicable from 1500 MHz to 2000 MHz and with a frequency correction factor f_c can be extended to higher frequencies. The path loss is calculated as

$$PL_{d,dB} = 46.2 + 33.9 \log_{10}(f) - 13.82 \log_{10}(T_h) - a(R_h)$$
$$+ [44.9 - 6.55 \log_{10}(T_h)] \log_{10}(d) + 0.7R_h + C + f_c, \quad (3.6)$$

where d is the path distance in kilometers, f is the frequency in megahertz, T_h is the base station (concentrator) antenna height, R_h is the subscriber station (DAP) antenna height, and f_c is calculated as:

$$f_c = \begin{cases} 26 \log_{10}(f/2000), & f > 2000 \, MHz \\ 0, & f \le 2000 \, MHz \end{cases} \quad (3.7)$$

For urban environments, $C = 3$ dB and

$$a(R_h) = 3.2[\log_{10}(11.75R_h)]^2 - 4.97. \quad (3.8)$$

For suburban environments, $C = 0$ dB, and

$$a(R_h) = R_h[1.1 \log_{10}(f) - 0.7] - [1.56 \log_{10}(f) - 0.8]. \tag{3.9}$$

The **Erceg model** is considered applicable from 1800 MHz to 2700 MHz; it also adopts a correction function for higher frequency. The path loss is calculated as:

$$PL_{d,dB} = 10(a - bT_h + c/T_h) \log_{10}(1000d/d_0)$$
$$+ 6 \log_{10}(f/2000) - X \log_{10}(R_h/2)$$
$$+ 20 \log_{10}(4\pi d_0 \lambda) + \delta(f). \tag{3.10}$$

In addition to the parameters used in the modified Hata model, here $d_0 = 100$ m, and λ is the wavelength in meters. The correction function $\delta(f)$ [71] for higher frequency is

$$\delta(f) = 10 \log_{10}(f/f_d)^{2+n}, n \leq 1, \tag{3.11}$$

where $f_d = 1900$ MHz and $n = 0.8$ in most settings. The Erceg model has three sets of settings which apply to three terrain types. Type A is hilly with moderate to heavy tree density, type B is hilly with light tree density or flat and moderate to heavy tree density, and type C is flat with light tree density. The detailed settings are shown in Table 3.2.

Suppose that the WiMAX operates at the 5.8 GHz unlicensed band, the path distance $d = 2$ km, the concentrator antenna height $T_h = 10$ m, and the DAP antenna height $R_h = 2$ m. The modified Hata model in a suburban environment gives $PL_{d,dB} = 143.44$ dB. As a comparison, the Erceg model [71] in hilly terrain with light tree density or flat terrain with moderate to heavy tree density gives 139.67 dB. The two models give close results; therefore, we adopt the modified Hata model for simplicity. Applying the maximum transmission power $p_{tx} = 30$ dBm allowed by the FCC, the receiving power is $p_{rx} = -113.44$ dBm respectively, which is higher than the

Table 3.2 The remaining parameters for the Erceg model.

Parameter	Type A	Type B	Type C
a	4.6	4.0	3.6
b	0.0075	0.0065	0.005
c	12.6	17.1	20
X	10.8	10.8	20

receiving sensitivity (widely accepted as −120 dBm). Apparently, operating at a lower licensed band will lead to lower path loss.

3.3 Problem Formulations and Solutions in the NAN Design

In this section, we will discuss the core problems of the proposed NAN design. The problems include *cost minimization, optimal number of gateways, deployment of gateway DAPs*, and *global uplink transmission power efficiency*. The problems and solutions explored in this section may serve as guidelines and benchmarks for future NAN deployment in the smart grid.

3.3.1 The Cost Minimization Problem

Generally speaking, the ultimate goal in the NAN design is to minimize the total cost for the NAN to operate for a certain number of years, for example, $m = 5$ years. The total cost of a NAN consists of three parts: deployment cost, operational cost, and maintenance cost. Deployment cost includes the cost of all the hardware, including DAPs, gateway DAPs, batteries, solar panels etc. Operation cost includes power usage (which is zero for renewable energy), the data subscription fee if a licensed band is used, etc. And maintenance cost includes the replacement of any malfunctioning equipment, plus labor costs and any monetary loss due to missing data.

In reality, the NAN is supposed to handle a small amount of data; thus it is safe to assume that a self-sustaining network without maintenance is more cost efficient. By adopting pure renewable energy and unlicensed bands, the operation cost is excluded as well. Therefore, it is reasonable to assume that the total cost for the NAN depends on the deployment cost only. In our NAN design, the equivalent problem is to find the *optimal combination of battery size and solar panel size for each DAP* so that the minimum total cost could be achieved. Let α be the price of a unit size battery (e.g. 1 Wh), and β be the price of a unit size solar panel. For the deployment, the objective is to find

$$\min \quad \alpha \sum_{k \in D} b_k + \beta \sum_{k \in D} s_k, \qquad (3.12)$$

where b_k is the battery size, s_k is the solar panel size for DAP k, and D is the set of all DAPs. For a DAP k to sustain the system for m years, its battery must last 365 m days before reaching the end of its state of health, that is, DoH_{dead}. Thus the battery is not supposed to be charged every day according to previous discussions. The battery should be charged only when its remaining energy falls below a threshold e_k^{th}. This threshold is set based on a critical situation, e.g. q continuous cloudy days, as follows:

$$e_k^{th} \geq q \left(\int_{t_r}^{t_r'} p_k(t)dt - \int_{t_r}^{t_s} \underline{c}_k(t)dt \right), \tag{3.13}$$

where $\underline{c}_k(t)$ is the charging rate for DAP k in cloudy weather and $p_k(t)$ is the power usage at time t. Let the original battery capacity of DAP k be E_k. For practical purposes, we also set the charging threshold to be no less than $0.2\,E_k$, such that

$$e_k^{th} \geq 0.2\,E_k. \tag{3.14}$$

Note that $e_k(t) < e_k^{th}$ will trigger the charging signal; however, the actual charging may not happen immediately. For example, the threshold may be reached during night time. Assuming that e_k^{th} is reached at t_s, then at sunrise time t_r', the remaining battery capacity will be $e_k^{th} - \int_{t_s}^{t_r'}$. In this case, if the next day is a sunny day, the battery should be fully charged at sunset t_s on that day. If the battery cannot be fully charged on a sunny day, it will not be fully charged under any circumstances. Because once $e_k(t) \geq e_k^{th}$, the indicator of charge is off, and the next charge cycle occurs when $e_k(t) < e_k^{th}$. However, the battery will not fully charged again, and the DAP consumes power for the whole day ($t_r \to t_r'$).

The sunny day charging constraint is set as follows:

$$\int_{t_r}^{t_s} \bar{c}_k(t)dt \geq E_k - e_k^{th} + \int_{t_r}^{t_r'} p_k(t)dt, \tag{3.15}$$

where $\bar{c}_k(t)dt$ is the charging rate in a sunny day at time t. Note that the daytime power consumption is merged with the previous nighttime power consumption and is expressed as the whole day power consumption. To make the illustration clearer, we also show in Eq. (3.16) that on a sunny day at sunrise time t_r, the charging rate is already above the corresponding power usage, such that

$$\bar{c}_k(t_r) \geq \max\{p_k(t)\}. \tag{3.16}$$

A larger solar panel is able to charge the battery faster; however, waste in resources needs to be avoided. Thus, let the upper bound be set such that the DAP can be self-sustaining if every day is cloudy throughout m years. Thus the constraint for the solar panel size is shown in Eq. (3.17).

$$\int_{t_r}^{t_s} \underline{c}_k(t)dt \le \int_{t_r}^{t'_r} p_k(t)dt. \tag{3.17}$$

Moreover, the battery must survive the designed life cycle, such that

$$DoH_k^{365m} \le DoH_{dead}. \tag{3.18}$$

Constraint Eq. (3.18) is further relaxed for simplicity and with the practical purpose of finally solving the optimization problem. We define τ_k as an estimate of the pseudo-shortest charging period for DAP k. It is estimated based on a fully charged DAP that operates through continuously cloudy days before it reaches the threshold. Thus τ_k is estimated as follows:

$$\tau_k = \left\lceil \frac{E_k - e_k^{th}}{\int_{t_s}^{t'_r} p_k(t)dt + \int_{t_r}^{t_s}[0, p_k(t) - c_k(t)]^+} \right\rceil, \tag{3.19}$$

where $[\cdot]^+$ returns the larger element of the two. The denominator is the remaining energy before a charging signal is triggered. The numerator is the DAP power consumption in a whole day, which consists of two parts. 1) $\int_{t_s}^{t'_r} p_k(t)dt$ is the nighttime power consumption, and 2) $\int_{t_r}^{t_s}[0, p_k(t) - c_k(t)]^+$ is the daytime power consumption. Obviously, if $c_k(t) \ge p_k(t)$, the DAP is powered by its solar panel and does not consume battery energy. Note that τ_k is not the shortest delay between two charges because a DAP with a partially charged battery that operates through continuous cloudy days may cause a shorter delay between two charges. However, if only extreme cases are considered, the design will end up with unnecessarily large battery and solar panel. The parameter τ_k is validated to be practical in the case study.

Let $\widetilde{Cy}_{max}$ be an estimated maximum charging cycle based on a deep DoD (e.g. 500 when $DoD = 80\%$). In order to survive the designed life cycle, instead of using constraint Eq. (3.18), we apply the following constraint:

$$\tau_k \ge \frac{365m}{\widetilde{Cy}_{max}}. \tag{3.20}$$

The DAP needs to record several types of data in two consecutive days (e.g. days i and $i + 1$) for further evaluation. First, it needs to record a time stamp t_c when $c_k(t_c) > p_k(t_c)$ for the first time and also record $e_k(t_c)$ on each day, (e.g. $e_k^i(t_c)$ for day i). The DAP also needs to record a daily charging cycle indicator (e.g. Cy_k^{i+1} for day $i + 1$). If $e_k^{i+1}(t_\mu) > e_k^{i+1}(t_c)$, then $Cy_k^{i+1} = Cy_k^i \oplus 1$, where $\oplus$ is an XOR function. Note that the battery will be charged at most once per day. Another indicator to be recorded is DoH. Based on Cy_k^i and Cy_k^{i+1}, DoH_k^{i+1} is calculated as follows:

$$DoH_k^{i+1} = \begin{cases} DoH_k^i, & Cy_k^{i+1} = Cy_k^i \\ \dfrac{DoH_k^{dead}}{Cy_{\max}^i} + DoH_k^i, & Cy_k^{i+1} \neq Cy_k^i \end{cases}, \tag{3.21}$$

where $Cy_{\max}^i$ is determined by $DoD_k^i = e_k^i(t_c)/E_k$ according to Eq. (3.5).

DAPs are independent of each other, as they do not share solar panels or batteries. Therefore, the total cost optimization problem can be approached individually for each DAP if the number of normal DAPs and gateway DAPs are determined (which will be discussed later). Therefore, solving the problem shown in Eq. (3.12) is equivalent to solving the problem as follows:

$$\min \; \alpha b_k + \beta s_k, \; \forall k \in D, \tag{3.22}$$

subject to
Constraints (3.19) and (3.20),

$$e_k^{th} \geq q \left\{ 24p_k - \frac{w_c}{2} \left[erf\left(\frac{t_s - \mu}{\sqrt{2\sigma_c^2}}\right) - erf\left(\frac{t_r - \mu}{\sqrt{2\sigma_c^2}}\right) \right] \right\}, \tag{3.23}$$

$$\frac{w_c}{2} \left[erf\left(\frac{t_s - \mu}{\sqrt{2\sigma_c^2}}\right) - erf\left(\frac{t_r - \mu}{\sqrt{2\sigma_c^2}}\right) \right] \leq 24p_k, \tag{3.24}$$

$$\frac{w_s}{2} \left[erf\left(\frac{t_s - \mu}{\sqrt{2\sigma_s^2}}\right) - erf\left(\frac{t_r - \mu}{\sqrt{2\sigma_s^2}}\right) \right] \geq E_k - e_k^{th} + 24p_k, \tag{3.25}$$

$$\frac{w_s}{\sqrt{2\pi\sigma_s^2}} e^{-\frac{t_r - \mu}{2\sigma_s^2}} \geq p_k. \tag{3.26}$$

Assume each DAP operates at a constant power consumption (i.e. $p_k(t) = p_k(1) = p_k \forall t$). After applying the charging rate estimation functions, new constraints Eq. (3.23), Eq. (3.24), Eq. (3.25), and Eq. (3.26) are rewritten from Eq. (3.13), Eq. (3.17), Eq. (3.15), and Eq. (3.16) respectively. With any given p_k, Eq. (3.22) is a linear optimization problem that can be solved efficiently. For normal DAPs, which are basically Wi-Fi routers, their power consumption is constant (e.g. $p = 100$ mW or 20 dBm). Therefore, the cost for normal DAPs can be determined so far. However, the power consumption for gateway DAPS ranges from 100 mW to 1 W. The method of deployment and number of gateway DAPs would highly affect the service quality to customers. It is not recommended to arbitrarily assign a power consumption level to a gateway DAP. Therefore, we need to first determine the number N of gateway DAPs and then optimize the uplink transmission power consumption p_k, $k \in \mathcal{N}$ for each of them.

3.3.2 Optimal Number of Gateways

In much of the research regarding power control for multiple-access systems [72–75], the effective transmission rate (or the receiving data rate) for a transmitter (e.g. gateway DAP k) is expressed as follows:

$$T_k = R_k f(\gamma_k), \tag{3.27}$$

where R_k is the physical layer transmission rate and γ_k is the signal-to-interference-plus-noise (SINR) ratio. The SINR is calculated as follows:

$$\gamma_k = \frac{p_k h_k}{\sigma^2 + \frac{1}{G} \sum_{j \neq k} p_j h_j}, \tag{3.28}$$

where G is the processing gain. In Eq. (3.27), $f(\gamma_k)$ is the efficiency function that calculates the probability of a packet being successfully transmitted. Function $f(\cdot)$ is widely adopted as an increasing, continuous, and S-shaped function with $f(\infty) = 1$, $f(0) = 0$ predetermined. Mathematically, the efficiency function can be as follows:

$$f(\gamma_k) = (1 - e^{-\gamma_k})^M, \ \gamma_k > 0, \tag{3.29}$$

where M is the packet length.

Let $\mathcal{N} = \{1, 2, \ldots, N\}$ be the indicators of gateway DAPs, and $N = |\mathcal{N}|$ be the total number of gateway DAPs. We then discuss the optimal N needed for a specific NAN. Since usage, coverage, and customer numbers of the smart grid, the optimal number is one that supports a maximum upload transmission data rate in a NAN that provides for future upgrades. This problem is formulated as follows:

$$N = \arg\max_{N \in \mathbf{N}^+} \sum_{k=1}^{N} T_k,$$ (3.30)

where $\mathbf{N}^+$ is the set of positive integers. We assume that each gateway DAP covers the same number of users regardless of its distance to the concentrator. Then the receiving power at the concentrator for each gateway needs to be identical, that is, $p_k h_k = p_{k+1} h_{k+1}$, $k \leq N - 1$, so that all smart meters will be treated fairly independent of their distance to the gateway. Since N shall be a constant for later discussion, we replace it with a variable n for solving Eq. (3.30) in this subsection. Eq. (3.28) is then rewritten as follows:

$$\tilde{\gamma}_k = \frac{p_k h_k}{\sigma^2 + \frac{n-1}{G} p_k h_k}.$$ (3.31)

For simplicity, path gain is represented as follows:

$$h_k = A d_k^{-\nu},$$ (3.32)

where d_k is the distance between the gateway k and the concentrator and ν is the fading parameter. Thus the receiving power can be written as

$$p_k^r = p_k^t h_k = A p_k^t d_k^{-\nu}.$$ (3.33)

Obviously, γ_k increases with respect to p_k^r, $\forall k \in \mathcal{N}$, $n > 1$. The maximum SINR $\gamma_{\max}$ is achieved when $p_k^r = p_{\max}^r$. Note that $p_{\max}^r$ depends on the furthest gateway, since it has the lowest path gain. Thus the maximum SINR is calculated as follows:

$$\gamma_{\max} = \frac{p_{\max}^r}{\sigma^2 + \frac{n-1}{G} p_{\max}^r}.$$ (3.34)

Note that the noise σ_2 (e.g. -120 dBm) is not negligible compared with $p_{\max}^r$ (e.g., -110 dBm). With a given $p_{\max}^r$, the efficiency function is a function with respect to $1/n$ (note that n is a positive integer) such that

$$f(\gamma_{\max}) \triangleq \tilde{f}(1/n).$$ (3.35)

The estimated maximum total transmission data rate from all the gateways $\tilde{T}_{total}$ is calculated as follows:

$$\tilde{T}_{total} = \sum_{k=1}^{n} T_k = n R_k \tilde{f}(1/n).$$ (3.36)

With that, the objective in Eq. (3.30) can be presented equivalently as follows:

$$N = \arg\max_{n \in \mathbb{N}^+} nR_k \tilde{f}(1/n). \tag{3.37}$$

Let $\tilde{n} \triangleq \frac{1}{n}$, then $\tilde{T}_{total}$ can be calculated equally as follows:

$$\tilde{T}_{total} = \frac{R_k \tilde{f}(\tilde{n})}{\tilde{n}}, \quad 0 < \tilde{n} < 1. \tag{3.38}$$

Theorem 3.1   Eq. (3.38) is quasiconcave with respect to $\tilde{n}$.

Proof: Because function $\tilde{f}(\tilde{n})$ is an increasing function with $\lim_{\tilde{n}\to 0} \tilde{f}(\tilde{n}) = 0$ and $\lim_{\tilde{n}\to 1} \tilde{f}(\tilde{n}) = 1$, function $\tilde{f}(\tilde{n})$, $0 < \tilde{n} < 1$ is an S-shape function, as shown in Figure 3.4. Therefore, $\tilde{f}(\tilde{n})/\tilde{n}$, $0 < \tilde{n} < 1$ is quasiconcave, as shown in Figure 3.5. And thus Theorem 3.1 holds.

□

Now that we know Eq. (3.38) is quasiconcave, it has a unique maximizer [76] $\tilde{n}^* = 1/n^*$ which can be found by calculating $\tilde{n}\tilde{f}'(\tilde{n}) = \tilde{f}(\tilde{n})$ [76]. To this end, the optimal number of gateways is found as follows:

$$N = \begin{cases} \left\lfloor \frac{1}{\tilde{n}^*} \right\rfloor, & \left\lfloor \frac{1}{\tilde{n}^*} \right\rfloor \tilde{f}\left(1 \Big/ \left\lfloor \frac{1}{\tilde{n}^*} \right\rfloor\right) \geq \left\lceil \frac{1}{\tilde{n}^*} \right\rceil \tilde{f}\left(1 \Big/ \left\lceil \frac{1}{\tilde{n}^*} \right\rceil\right) \\ \left\lceil \frac{1}{\tilde{n}^*} \right\rceil, & \left\lfloor \frac{1}{\tilde{n}^*} \right\rfloor \tilde{f}\left(1 \Big/ \left\lfloor \frac{1}{\tilde{n}^*} \right\rfloor\right) < \left\lceil \frac{1}{\tilde{n}^*} \right\rceil \tilde{f}\left(1 \Big/ \left\lceil \frac{1}{\tilde{n}^*} \right\rceil\right) \end{cases}, \tag{3.39}$$

where $\tilde{n}^* = \arg_{\tilde{n}} \tilde{n}\tilde{f}'(\tilde{n}) = \tilde{f}(\tilde{n})$.

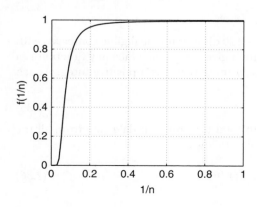

Figure 3.4 S-function $f(1/n)$.

Figure 3.5 A quasiconcave function.

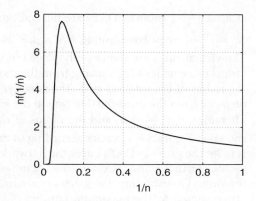

3.3.3 Geographical Deployment Problem for Gateway DAPs

Intuitively, if the gateway DAPs are deployed close to the concentrator, they would have the best network performance. However, metering data from those customers who are further away from the concentrator will go through more hops before finally reaching the concentrator and thus incur longer delays, which decrease the quality of service provided to them. Therefore, the purpose of the geographical deployment of gateway DAPs is to let all the customers get high-quality service.

Without loss of generality, the neighborhood is assumed to be a square block with side length d. Houses are usually built in a grid topology such that the smart meters are close to uniform distribution. DAPs also need to be deployed as uniformly as possible so that each customer can be fairly served. The fairness also applies to gateway DAPs. Achieving exact fairness for all customers is beyond the scope of this work, and it can hardly be done considering solely the geographical deployment. We then propose a method to approach some geographical fairness. Further fairness can be achieved through the routing scheme in the local multihop transmission.

In an area that covers $S = d^2$ (e.g. 1 km^2), N gateway DAPs are to be deployed (or chosen from existing DAPs). Each gateway DAP shall cover an area of S/N in order to be fair. Suppose the coverage of a gateway is a disk shape, then the radius of an individual disk is $r_0 = \sqrt{\frac{S}{\pi N}}$. Let the gateway DAPs be deployed on a set of virtual rings which are centered at the concentrator. The number of rings N_r depends on r_0 and d. Due to the symmetric nature of the topology, we only need to process a half of it (e.g. the right half). Within $0.5d$, at least $\lceil \frac{0.5d}{2r_0} \rceil$ gateway DAPs are needed horizontally. That indicates $N_r = \lceil \frac{0.5d}{2r_0} \rceil$ rings. For example, if $N = 3, 8, 24, N_r = 1, 2, 3$ respectively. The virtual rings are then evenly placed in the neighborhood with a radius increment of $2r_0'$ starting from the most inside one, where r_0' is the radius of the smallest ring. The smallest radius is also the closest distance from the most outside ring to the edge of the neighborhood. Therefore, $r_0' = \frac{0.5d}{2N_r} \times 2$, and the radius of ring i is $r_i = (2i - 1)r_0'$. As shown in Figure 3.6, r_0' varies according to different N_r. The idea here is to let the gateway DAPs cover the network as evenly as possible.

Once virtual rings are established, gateways can be distributed on the rings. On each ring, the gateways are roughly $2r_0$ away from each other; therefore, the maximum number of gateways on a ring i is as follows:

$$N_r^i = \left\lceil \frac{2\pi r_i}{2r_0} \right\rceil = \left\lceil \frac{\pi r_i}{r_0} \right\rceil$$

One may have realized that not all gateways are guaranteed to be deployed on the virtual rings evenly. Thus we determine the number of gateways on each ring first and then deploy them on each ring. Let $\mathbf{d} = \{d_1, d_2, \ldots, d_N\}$ be the set of distances from gateways to the concentrator. Each distance d_i reveals the ring that gateway i is deployed. We propose two methods to find the set of distances $\mathbf{d}$.

In the first method, gateways are deployed one on each ring, starting from the outermost ring. The maximum number of gateways on each ring is determined by N_r^i. Using $N = 8$ and $d = 1$ km, the deployment is illustrated in Figure 3.7. The gateway DAPs are deployed along two virtual rings, where the outer ring has six and the inner ring has two.

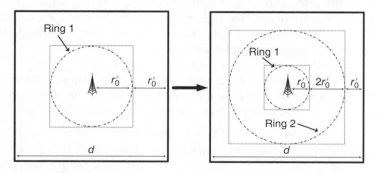

Figure 3.6 Illustration of rings.

Figure 3.7 Illustration of "one on each" method.

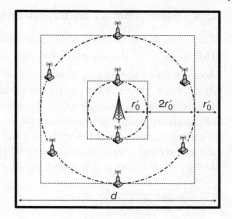

In the second method, each ring is fully deployed on a virtual ring before moving onto the next ring, starting from the outermost ring. As shown in Figure 3.8, the deployment is different from that using the first method. While the gateway DAPs are also deployed along two virtual rings, the second method deploys just one on the inner ring. Apparently, the outsider ring first method will provide a better level of fairness to the customers that are further away from the concentrator. However, more gateways on the outsider rings have side effects on the uplink transmission performance, which will be discussed later in this chapter.

Figure 3.8 Illustration of "outsider ring first" method.

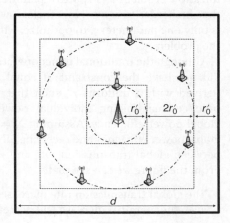

3.3.4 Global Uplink Transmission Power Efficiency

Many researches have considered individual power efficiency (e.g. for DAP k, $u_k = \frac{T_k}{p_k^t}$ (bits/joule)) for each node in a multiaccess system. Most of them applied noncooperative game theoretical approaches where each player achieves its maximum power efficiency selfishly. However, that approach cannot achieve the maximum social welfare (or the global uplink transmission power efficiency). The global power efficiency is defined as the ratio of total effective uplink transmission rate to the total transmission power. With a given topology for gateway DAPs, the problem is described as follows:

$$\max \frac{\sum_{k \in \mathcal{N}} T_k}{\sum_{k \in \mathcal{N}} p_k^t}, \tag{3.40}$$

such that

$$p_k^t \leq p_{\max}^t, \ \forall k \in \mathcal{N}, \tag{3.41}$$

and

$$\gamma_k = \gamma_{k+1}, \ \forall k \in \mathcal{N}. \tag{3.42}$$

Constraint Eq. (3.41) indicates that the transmitting power p_k^t for each gateway has an upper bound imposed by the FCC rules (especially using ISM bands). Constraint Eq. (3.42) is for the fairness. As discussed before, all gateway DAPs shall have the same receiving power at the concentrator. Without this fairness constraint, this nonlinear optimization problem has N extra parameters (γ_k, $\forall k \in \mathcal{N}$) and would thus be harder to solve. Nonetheless, with fairness introduced, there is only one parameter γ to optimize, which reduces the complexity of the problem.

Although the traditional noncooperative game theoretical approach did not have the constraint of equal γ, its Nash equilibrium was derived with a uniform γ^* such that $\gamma^* f'(\gamma^*) = f(\gamma^*)$ [72–75]. In practice, maximizing individual power efficiency cannot achieve global power efficiency. Assuming $N = 18$, the relationship between global power efficiency and receiving SINR γ as well as the relationship between global transmission rate and γ are illustrated in Figure 3.9. From the figure we can see that:

- The global transmission rate increases with respect to SINR. However, the marginal increase will decrease as SINR goes higher.

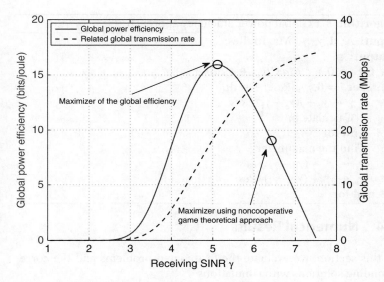

Figure 3.9 Global uplink transmission power efficiency.

- A maximizer of the global power efficiency exists. The maximizer does not necessarily represent the optimal SINR that can support the data transmission rate.
- The maximizer achieved by noncooperative game theoretical approach will be off from the global maximizer. It requires a higher SINR, which benefits the transmission rate.

To find optimal solution to global power efficiency, we propose an algorithm as illustrated in Algorithm 3.1. Let $\mathbf{p} = \{p_1, p_2, \ldots, p_N\}$ be the set of power usages of each DAP. Let *MaxItr* be the number of iterations, and *Res* be the resolution of each iteration. In each iteration starting from γ_s and ending at γ_e, there is one maximizer γ^* to be found. In the next iteration, the algorithm updates $\gamma_s \leftarrow \gamma^* - (\gamma_e - \gamma_s)/Res$ and $\gamma_e \leftarrow \gamma^* + (\gamma_e - \gamma_s)/Res$. With a given γ, the computational complexity depends on calculating $\mathbf{p}$. Fortunately, calculating $\mathbf{p}$ is indeed solving a linear equation system of N linear equations with N unknowns. Its computational complexity has an upper bound at $O(N^{\log_2 7})$ when applying Strassen's algorithm [77] for matrix multiplication. Therefore, Algorithm 3.1 has an upper bound for computational complexity $O(\kappa N^{\log_2 7})$, where $\kappa = MaxItr \times Res$.

Algorithm 3.1 Global power efficiency approach

Input: N, **d**, γ_s, γ_e, MaxItr, Res;

Output: **p**;

  1: **for** $i = 0, i < MaxItr, i + +$ **do**
  2:   **for** $j = 0, j < Res, j + +$ **do**
  3:     $\gamma \leftarrow \gamma_s + j(\gamma_e - \gamma_s)/Res$;
  4:     Calculate **p**;
  5:   **end for**
  6:   Find the maximizer γ^*;
  7:   $\gamma_s \leftarrow \gamma^* - (\gamma_e - \gamma_s)/Res$
  8:   $\gamma_e \leftarrow \gamma^* + (\gamma_e - \gamma_s)/Res$
  9: **end for**

3.4 Numerical Results

In this section, we evaluate the formulated problems and the corresponding solutions with simulations.

3.4.1 Evaluation of the Optimal Number of Gateways

Let the noise $\sigma^2 = -120$ dBm, M=100 bits, and $G = 128$. Assume a neighborhood block with $d = 1$ km. We first show the maximum number of gateway DAPs with respect to different maximum receiving power p^r_{max}. Given a p^r_{max}, an optimal number of gateway DAPs can be found by the solution given in the previous section. As shown in Figure 3.10, with higher p^r_{max}, the optimal N is higher. However, as discussed before, deploying more gateway DAPs causes lower p^r_{max} since we need to achieve geographical fairness. Therefore, it cannot deploy too many gateway DAPs while applying low p^r_{max}.

3.4.2 Evaluation of the Global Power Efficiency

The global power efficiency depends on the number as well as the deployment of gateway DAPs. The proposed centralized optimization and the traditional game theoretical approach may achieve different results. In the evaluation, geographical issues are considered is only for the centralized approach, since the game theoretical approach is independent of actual deployment as long as the concentrator can still receive from the furthest gateway DAP at optimal SINR.

Figure 3.11 shows the global power efficiency in bits/joule. In general, the global power efficiency decreases as the number of

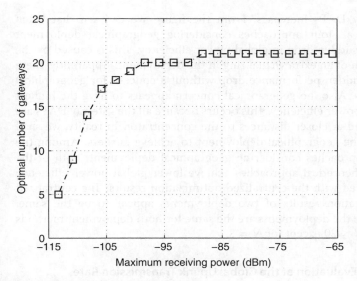

Figure 3.10 Optimal number of gateways.

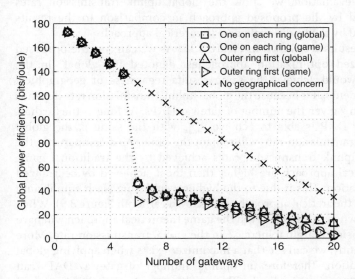

Figure 3.11 Global power efficiency with respect to number of gateways.

gateway DAPs increases. From the figure we can see that when $N = 7$, all four approaches considering geographical deployment have a sudden drop in global power efficiency. This is caused by the addition of an extra virtual ring in the deployment. Apparently, there is no sudden performance drop without a concern for geographical fairness. Also, no geographical concern appears to have the highest global power efficiency. This occurs because all the gateway DAPs are deployed at closer distances to the concentrator. In reality, we shall adopt the geographical deployment to achieve fairness. Among the four approaches considering geographical deployment, both of the game theoretical approaches achieve lower global power efficiency compared with the centralized optimization results. The centralized optimization results of two deployments appear to be the same, because the deployments are the same for both deployment methods when $N \leq 20$ except for $N = 8$.

3.4.3 Evaluation of the Global Uplink Transmission Rates

In this evaluation, we show the global uplink transmission rates achieved by the proposed approach in comparison to the results achieved by the traditional game theoretical approach.

The results are shown in Figure 3.12. As we can see, the proposed centralized optimization approaches, denoted by "global" in the figure, would achieve the same results regardless of geographical deployments. It is because the global uplink transmission rate depends solely on γ given the number of gateway DAPs. As long as the furthest gateway DAP is able to achieve p_{max}^{r} with the same γ, the global uplink transmission rate will remain the same. As a comparison, the global uplink transmission rates achieved by the traditional game theoretical approach are higher than those achieved by centralized optimization. This is due to the higher γ achieving Nash equilibrium in game theoretical approaches (as shown earlier in Figure 3.9). While not shown in this figure, even the game theoretical approach will have a decreasing marginal increase in the global transmission rate. More importantly, we can see that a maximizer exists when applying global optimization. Therefore, an optimal number of gateway DAPs can be determined accordingly. The results also show that if increased service is needed for more customers, the NAN uplink transmission performance can be enhanced easily by increasing transmission power of each gateway DAP. There is no need for a hardware upgrade in this structure.

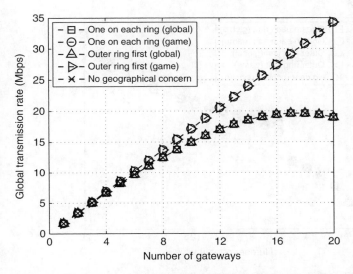

Figure 3.12 Global transmission rate with respect to number of gateways.

3.4.4 Evaluation of the Global Power Consumption

In this evaluation, we show the global power consumption of the gateway DAPs with different approaches. The results are given in Figure 3.13.

We can see that with no geographical consideration, the global power consumption is the lowest. This is because when no deployment issue is considered, the gateway DAPs are actually deployed closer to the concentrator, resulting in lower power consumption. Moreover, the game theoretical approaches have slight higher global power consumption due to their correspondingly higher γ.

3.4.5 Evaluation of the Minimum Cost Problem

The solution to the minimum cost problem is to find the sufficient size of solar panel and capacity of battery for each DAP to survive the designed life cycle. The transmission power p_k of a gateway DAP is determined by solving the maximum global power efficiency problem after finding the optimal number of gateway DAPs. The transmission power of a normal DAP is a constant value set for Wi-Fi transmissions. With a given transmission power p_k, the key factors of the minimum cost problem are the length of life cycle m as well as

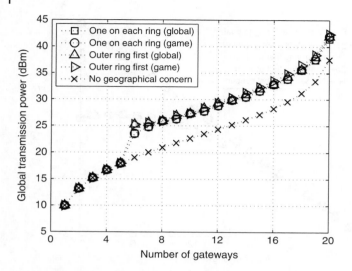

Figure 3.13 Global power consumption with respect to number of gateways.

the estimated worst-case scenario q. This evaluation is to show the relationships between those factors. Without loss of generality, let $\widetilde{Cy}_{max} = 500$, $\alpha = \$2$ per Wh, and $\beta = \$6$ per unit solar panel for the maximum 1 W charging rate. With $q = 3$, evaluation is performed with $m = 1, 2, 3, 4, 5$. In addition to that, when $m = 5$, evaluation is also performed with $q = 1, 2$. Different settings can be easily applied to set other benchmark evaluations.

First, we evaluate the impact on total cost (the objective function) with respect to different values of p_k. The results are given in Figure 3.14. As we can see, either higher m or higher q will increase total cost for a DAP, since it will need a larger battery and larger solar panel to be self-sustaining over a longer and harsher life cycle.

Then, we evaluate the impact on battery capacity with respect to different values of p_k. The results are given in Figure 3.15. As we can see, a larger battery is required with a longer life cycle and higher transmission power.

The impact on solar panel size is shown in Figure 3.16. Note that the solar panel size is the same when $m = 5$ and $q = 1, 2, 3$. Although the battery capacities are different, the charging rate is high enough to handle the situations.

Finally, we show the impact on charging threshold in Figure 3.17. As it shows, with a higher q, the charging threshold increases. However,

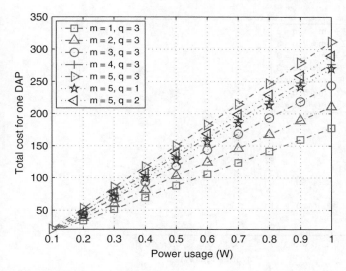

Figure 3.14 Total cost of a DAP.

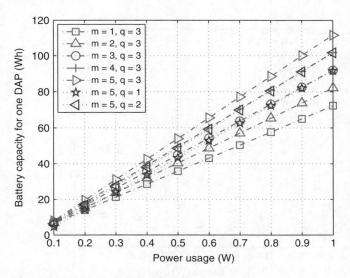

Figure 3.15 Battery capacity of a DAP.

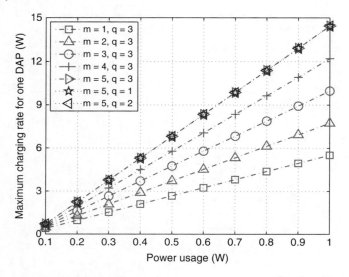

Figure 3.16 Solar panel size of a DAP.

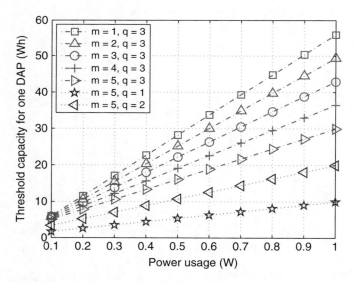

Figure 3.17 Impact on charging thresholds of a DAP.

with a higher m, the charging threshold decreases. On one hand, more capacity remaining in the battery is required to survive a longer period with bad weather. On the other hand, if the required life cycle is longer, then the battery must reduce its charging frequency to slow down the deterioration of that battery.

3.5 Case Study

In this section, we present a case study to demonstrate the self-sustainability of the proposed NAN design with the results from previous evaluations. In the case study, the neighborhood has a dimension of d=1 km. The optimal number of gateway DAPs in this neighborhood is 18. Both of the deployment methods will have the gateway DAPs placed in the same place. The furthest gateway DAP (i.e. DAP k) operates at $p_k = 304$ mW to achieve maximum global power efficiency. Assume that the NAN is designed for $m = 5$ years and $q = 3$ *days*. The optimal settings of each battery and solar panel are $E_k = 31.01$ *Wh*, $w_k^s = 19.15\ w_0^s$, and $e_k^{th} = 10.62$ *Wh*. The weather conditions are assumed to be random throughout a year but remain constant in a day for simplicity. The NAN is evaluated with different probabilities of sunny weather.

First, we show the remaining energy at sunrise on each day in Figure 3.18 with a 20% probability of sunny weather. From the results, we can see that the exact battery capacity drops below the original capacity as time passes due to deterioration. It also shows that no charging begins when $E_k \geq e_k^{th}$; however, charging may not begin not only below the charging threshold, but also when $E_k < 0.2E_k$ due to continuous cloudy days. Sometimes, the battery could be charged just before it gets completely depleted. One may use a slightly larger capacity or higher charging threshold to enhance self-sustainability.

A closer look at *DoH* is shown in Figure 3.19. With a higher probability of sunny weather, the *DoH* is lower. A simple explanation is that a larger number of sunny days makes the DAP more likely to activate a charger right after the remaining energy hits the threshold and thus causes more charging cycles. Although the *DoH* after each charge depends on the corresponding *DoD*, the *DoDs* in this case are more than 65.75% when Cy_{max} is close to 500. Therefore the approximation is reliable. Note that the final *DoHs* of all situations are below

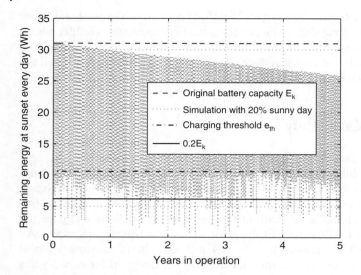

Figure 3.18 Remaining energy at t_s every day.

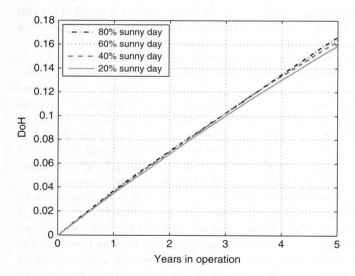

Figure 3.19 *DoH* at t_r every day.

$DoH_k^{dead} = 0.2$, and thus this battery will survive throughout five years. Different settings can be applied before actual deployment of a NAN for better self-sustainability.

3.6 Summary

In this chapter, we proposed an energy-efficient, self-sustaining wireless NAN design that is powered by a solar panel. We proposed a centralized optimization method so that the cost of the NAN can be minimized. In order to find the solution, we discussed the optimal number of gateway DAPs. We also applied two geographical deployment algorithms so that the fairness of high-quality service can be achieved for all the customers, especially ones farther from the concentrator. When determining the transmission power of each gateway DAP, we tailored the approach for global power efficiency instead of individual power efficiency. Compared with the traditional game theoretical approach, our proposed approach can achieve lower power usage and higher power efficiency. The case study demonstrated that our design is self-sustaining.

4

Reliable Energy-Efficient Uplink Transmission Power Control Scheme in NAN

In the previous chapter, we proposed a self-sustaining wireless neighborhood-area network (NAN) design. In this chapter, we propose a hierarchical power control scheme for the uplink transmission targeting the real-time data rate reliability and energy efficiency. A two-level Stackelberg game is applied to the proposed scheme, and the Stackelberg equilibrium is derived, based on a linear receiver. The proposed power control scheme is demonstrated to achieve both data rate reliability and energy efficiency.

4.1 Background and Related Work

4.1.1 Motivations and Background

Our proposed NAN design in the previous chapter was based on fixed data rate requirements in the smart grid. In practice, the data rate requirement may not be that demanding, since the smart grid will be rolled out in phases. Therefore, a NAN does not need to operate at it full designed capacity in early phases. In our proposed wireless NAN design, the networks are enabled by Wi-Fi and WiMAX. Normal data aggregate points (DAPs) are enabled with Wi-Fi for internal communications in a NAN. Gateway DAPs are also enabled with WiMAX for bridging a NAN and the utility backbone network.

WiMAX is chosen for gateway DAPs because it may be deployed in in the unlicensed 5.8 GHz unlicensed spectrum. Avoiding licensing fees could significantly lower costs. However, for systems operating in unlicensed bands, the maximum effective isotropic radiated power (EIRP) is limited to 30 dBm (or 1000 mW) by the Federal

Smart Grid Communication Infrastructures: Big Data, Cloud Computing, and Security,
First Edition. Feng Ye, Yi Qian, and Rose Qingyang Hu.
© 2018 John Wiley & Sons Ltd. Published 2018 by John Wiley & Sons Ltd.

Communication Commission (FCC) [61]. If the equipment transmits in a fixed point-to-point link, the maximum EIRP can be slightly higher. Thus in our proposed network model, gateway DAPs are equipped with point-to-point WiMAX technology operating in 5.8 GHz bands to take advantage of the free spectrum and to achieve better performance. Compared to a normal Wi-Fi based DAP, which operates at 100 mW, the power consumption for each gateway is a more important issue. If a gateway DAP can operate for a longer life cycle, so can a normal Wi-Fi-based DAP.

Increasing the energy efficiency of a NAN benefits its longevity as well as self-sustainability. One needs to realize that one purpose of a NAN in the smart grid is to deliver information such as metering data. The advanced metering infrastructure (AMI) requires reliable data delivery to enable functions in the smart grid. In order to analyze the performance of a NAN transmission, we define *data rate reliability* as the ratio of the total effective uplink transmission data rate to the data rate generated by all the smart meters in the NAN. Data rate reliability is a metric to measure if a NAN successfully uploads sufficient smart meter data to the concentrator.

In this chapter, we focus our study on the uplink transmission, because the downlink transmission contains only a relatively small amount of information from the metering data management system (MDMS). Unlike the downlink transmission, achieving a very high data rate reliability (e.g. 99.999%) for the uplink transmission is challenging because of the increasing real-time transmission data rate requirement as the smart grid rolls out. With more pervasive deployment of the smart grid in near future, smart meter data will be generated much more frequently than the current 15-minute-based sampling period. For example, if we adopt currently used 12 KB as one sampling smart meter data, and the minimum sampling period goes down from 15 minutes to 1 second, the data rate required by a smart meter would increase almost 100 times, from 100 bps to 96,000 bps. Moreover, it is estimated that a yearly amount data as massive as 100 PBytes will be generated by AMI within 10 years [78]. This will quickly cause storage and processing problems for the MDMS. Therefore, if a smart meter is *smart* enough to adaptively change its sampling frequency based on the household situation, the peak volume of smart meter data can be lowered dramatically, and this in turn reduces the burden of the MDMS. For example, fewer people stay at home during daytime on working days lower metering frequency;

more appliances are used from evening to midnight when people are at home thus higher metering frequency.

4.1.2 Related Work

The uplink transmission from the gateway DAPs to the concentrator occurs over a multiaccess wireless network. There has been much research about the power efficiency for this topic [72–75]. However, that research was based on the noncooperative game theoretical approach, which achieves individual optimality. As an alternative, our proposed scheme is intended to achieve global optimality. Moreover, fairness to the customers is considered in our NAN design, so even the furthest customer in the neighborhood can receive the same quality of service provided by the smart grid communication infrastructure. As discussed earlier in the previous chapter, we consider geographical fairness when deploying gateway DAPs so that each gateway would roughly cover the same number of customers. Moreover, gateway DAPs operate at different power levels so that the receiving power at the receiver side is uniform. Besides fairness, it is also a requirement to consider the geographical deployment of gateway DAPs, since it is the first priority of a NAN to upload sensing/measuring status data from even the most distant customer to the concentrator within the latency requirement.

A wireless mesh network based on the IEEE 802.11s protocol [79] has been widely used in the literature as one of the best candidates for deploying a NAN [58, 63, 80]. A Wi-Fi-enabled NAN can handle high data rate transmission within the network, owing to the robust and still improving Wi-Fi technology. Some uplink transmission issues have been studied, such as the uplink transmission of AMI in terms of latency and throughput based on multigate structure [63, 80]. However, energy efficiency was not mentioned in these works as they did not address green energy, which plays an important role for the power supply of the AMI in the smart grid. The deployment of an AMI with green energy (e.g. solar panel and battery) was studied in [10] without a specific power control scheme to further enhance the performance of a NAN. Energy efficiency for a multiple-access system has been proposed in [72, 74, 75] by achieving maximum bits/joule utility. Since AMI is distinguished from those general multiple-access systems by its demanding data rate and low latency requirement, the results from those studies cannot be directly applied to AMI.

In the rest of this chapter, we propose an uplink transmission power control scheme that aims to guarantee data rate reliability while maintaining a high level of energy efficiency in a NAN. Specifically, the proposed scheme for uplink transmission power control is based on a two-level Stackelberg game theoretical approach.

4.2 System Model

The general system model follows the proposed NAN structure in the previous chapter. Without loss of generality, DAPs are further grouped into dual-link gateway DAPs (DGDs), basic gateway DAPs (BGDs), and normal DAPs, as illustrated in Figure 4.1.

All DAPs are equipped with Wi-Fi capability. In addition, BGDs are equipped with an extra one-way point-to-point WiMAX interface. DGDs are equipped with an extra two-way point-to-point WiMAX interface. Both DGDs and BGDs are responsible for the uplink transmission of the NAN, while only DGDs are enabled for downlink transmissions. This design has two reasons: 1) Since the downlink transmission data rate is low and is far less frequent, it is unnecessary to enable all the gateways, and 2) the concentrator can also use point-to-point transmission with limited transmission power as required by FCC regulations if the number of receivers is small. All gateway DAPs serve as the root in the hybrid wireless mesh protocol (HWMP) [79] in a NAN, so that metering data are transmitted to any of the gateways according to the protocol. If a gateway is not able to

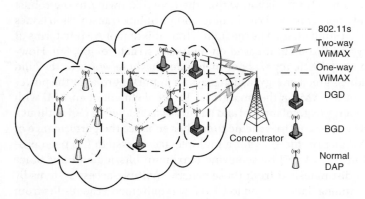

Figure 4.1 Illustration of the studied NAN structure.

handle its current incoming data rate alone, it will forward the data to other gateways. For simplicity, we ignore the interference among the gateways due to the nature of directional transmissions.

The power control scheme is for uplink transmission from a NAN to the concentrator. Downlink transmission is not considered in the scheme because downlink transmissions contain only a small amount of control information from the MDMS. It is transmitted much less frequently (e.g. once in an hour). However, metering data is generated as a massive continuous data flow and is delay sensitive. In an uplink transmission, we assume all the gateway DAPs simultaneously upload data toward one concentrator over the block Rayleigh flat-fading channels. All the uplink channel state information (CSI) on each block is perfectly known by the concentrator. Each gateway DAP knows only its own CSI. The receiving signal at the concentrator can be mathematically expressed as follows:

$$Y = \sum_{k=1}^{K} h_k X_k + Z, \qquad (4.1)$$

where Z is a zero-mean white Gaussian noise with variance σ^2, h_k is the fading channel gain, and X_k is the input signal of transmitter k [74].

Instead of considering the amount of data generated by smart meters, we consider its corresponding real-time data rate. The total incoming real-time data rate at the gateways is equivalent to the required uplink transmission data rate. For simplicity, we assume that no data loss is tolerable in the system. When each smart meter generates data at its minimum sampling period (e.g. 1 sec), the total incoming real-time data rate is at its maximum value $R_{\max}^S$. The number of DGDs and BGDs are determined by $R_{\max}^S$ before establishing the NAN. More specifically, DGDs are designed to handle a portion of the total data, that is, $\alpha R_{\max}^S$, $0 \leq \alpha \leq 1$ (without considering BGDs). With BGDs included, all the gateways handle all data uplink transmissions, that is, $(\alpha + \beta)R_{\max}^S$, $0 \leq (\alpha + \beta) \leq 1$.

4.3 Preliminaries

Some preliminaries are presented in this section, including a mathematical model and an energy-efficiency function that are to be applied to the uplink transmission power control scheme.

4.3.1 Mathematical Formulation

Table 4.1 lists the key notations and terminology used throughout the rest of the section. We denote the DGD set as $\mathcal{D} = \{d_1, \ldots, d_D\}$ and the BGD set as $\mathcal{B} = \{b_1, \ldots, b_B\}$. The numbers of DGDs and BGDs are $D = |\mathcal{D}|$ and $B = |\mathcal{B}|$ respectively. Note that if $B = 0$, then all gateway DAPs support two-way communication.

Let λ_k^D be the incoming data rate to DGD d_k, and λ_k^B be the incoming data rate to BGD b_k. The total data rate required by the smart meters is calculated as follows:

$$R_{in}^S = \sum_{i=1}^{D} \lambda_i^D + \sum_{i=1}^{B} \lambda_k^B. \tag{4.2}$$

Unlike the packet loss rate that only considers the data rate of the transmitters, the data rate reliability is a metric to measure whether or not the NAN successfully uploads sufficient smart meter data to the concentrator. For example, $\eta = 99.999\%$ indicates that all the transmitters together should have an effective uplink transmission data rate over 99.999% of the incoming data rate. However, it does not require the total packet loss rate to be less than 0.001% when each

Table 4.1 Key notations and terminology.

Sets	
$\mathcal{D}$	set of DGDs
$\mathcal{B}$	set of BGDs
$\mathbf{P}_D$	set of transmitting power of $d_k, \forall k \in \mathcal{D}$
$\mathbf{P}_B$	set of transmitting power of $b_k, \forall k \in \mathcal{B}$

Variables	
λ_k^D	incoming data rate to d_k
λ_k^B	incoming data rate to b_k
R_{in}^S	total generated uplink data rate
R_{out}^S	total transmitted uplink data rate
p_k^D	transmitting power of d_k
p_k^B	transmitting power of b_k
γ_k^D	output SINR for d_k
γ_k^B	output SINR for b_k

transmitter has a relatively high transmission rate compared to its incoming data rate.

Definition 4.1 *Data Rate Reliability* Letting R_{out} be the total uplink transmission data rate to a NAN (will be specified later), we define the data rate reliability η of a NAN as

$$\eta = \min\left(\frac{R_{out}}{R_{in}^S}, 1\right).$$

4.3.2 Energy Efficiency Utility Function

It is known that additional energy consumption changes the fundamental tradeoff between energy efficiency and the data rate [81]. Intuitively, achieving a higher signal-to-interference-plus-noise ratio (SINR) level requires the user terminal to transmit at higher power, which would result in lower energy efficiency. This tradeoff is well known and can be quantified [72–74]. The pure *energy efficiency utility function* of a user is defined as the ratio of its throughput to its transmit power (bits/joule) as follows:

$$u = \frac{T_k}{p_k}, \tag{4.3}$$

where T_k is the receiving data rate calculated as follows:

$$T_k = R_k f_e(\gamma_k). \tag{4.4}$$

Function $f_e(\gamma_k)$ is the efficiency function, which is assumed to be increasing, continuous, and S-shaped (sigmodial; more specifically, there is a point above which the function is concave, and below which the function is convex [76]) with $f_e(\infty) = 1$ and $f(0) = 0$. This efficiency function is commonly adopted by many researchers [73–75] as follows:

$$f_e(x) = (1 - e^{-x})^M, \tag{4.5}$$

where M is the block length and γ_k is the output SINR for k-th gateway. Assuming random spreading sequences, the SINR is calculated as follows:

$$\gamma_k = \frac{p_k h_k}{\sigma^2 + \frac{1}{N}\sum_{j\neq k} p_j h_j}, \tag{4.6}$$

where N is the processing gain, and p_k and h_k are the transmit power and path gain of the k-th gateway respectively. Eq. 4.6 can be rewritten to calculate the transmit power as follows:

$$p_k = \frac{\gamma_k(\sigma^2 + \frac{1}{N} \sum_{j \neq k} p_j h_j)}{h_k} \tag{4.7}$$

According to Eq. 4.4, the total transmitted uplink data rate R_{out} is calculated as follows:

$$R_{out} = \sum_{i=1}^{D} R_i^D f_e(\gamma_i^D) + \sum_{i=1}^{B} R_i^B f_e(\gamma_i^B). \tag{4.8}$$

With a given SINR, the energy utility function is quasiconcave. An illustration of the characteristics of a quasiconcave function is given in Figure 4.2, as indicated by the solid line. As it shows, there is a maximizer that can achieve the optimal pure energy efficiency Eq. (4.3) with a given SINR γ_k. However, the maximizer p_k^* may not guarantee a required data transmission rate. According to the efficiency function Eq. (4.5), the higher the transmit power, the higher the data rate. Therefore, when the total generated uplink data rate R_{in}^S is high, transmitting with p_k^* is more likely to miss the data rate reliability requirement.

In order to meet the data rate reliability η requirement in all scenarios while maintaining high energy efficiency, we introduce a weighted

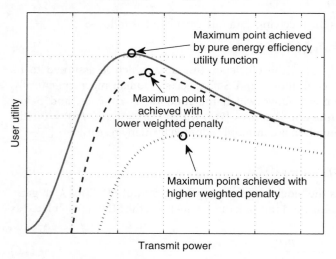

Figure 4.2 Illustration of utility function with/without penalty.

penalty function, denoted as $r_k(\cdot)$. The penalty serves as a measure of the reliability gap caused by insufficient transmitting power. For practical purposes, we require $r_k(\cdot)$ to satisfy three properties with respect to p_k with given noise:

1) It is a decreasing function.
2) It is convex.
3) It approaches 0 at infinity.

With $r_k(\cdot)$, we then define the utility of the k-th gateway DAP as follows:

$$u_k = T_k/p_k - r_k. \tag{4.9}$$

For example, when a gateway DAP (e.g. k) transmits at a higher power than the maximizer of the pure energy efficiency, the penalty is less, because transmitting at higher power level would be more likely to meet the reliability requirement. The convexity of $r_k(\cdot)$ ensures that the weight of the penalty as the power consumption increases. In the defined energy efficiency, the marginal increase in data rate decreases as transmit power increases higher than the maximizer; thus the penalty should follow this characteristic. The penalty approaches zero when consuming infinite power.

The impacts of the penalty function are also illustrated in Figure 4.2. The maximizer calculated based on the utility function Eq. (4.9) is used as a benchmark point. The penalty in the rewritten utility Eq. (4.9) shifts the maximizer value p_k to the right of the benchmark value. With a higher weighted penalty, the maximizer would be shifted more to the right. As shown in Figure 4.2, the maximizer of the dotted line is shifted higher than the maximizer of the dashed line, which means the penalty is weighted more in the dotted line. The exact penalty function will be presented with the transmission power control scheme in the next section.

4.4 Hierarchical Uplink Transmission Power Control Scheme

In this section, we present the hierarchical uplink transmission power control scheme for DGDs and BGDs. We propose to use a two-level Stackelberg game theoretical approach, where the DGDs act as the leader and the BGDs act as the follower. Generally speaking, the DGDs play a noncooperative game by knowing the reaction function of the BGDs to their actions. With the action of DGDs, BGDs play a noncooperative game among each other.

4.4.1 DGD Level Game

The DGD level game is defined as $\mathcal{G}_D = [D, \{p_i^D\}, \{u_i^D(\cdot)\}]$. In this level of game, $D = (d_1, \ldots, d_D)$ are the players, that is DGDs; $\{p_i^D\}$ is the action set (i.e. power consumption of each DGD), and $\{u_i^D(\cdot)\}$ is the set of individual utilities (i.e. power efficiency with defined penalty).

Let $p_{\max}^D$ be the maximum transmitting power for all the players. For an arbitrary player d_k, given its action $p_k^D \in (0, p_{\max}^D]$, the corresponding utility is calculated as follows:

$$u_k^D = T_k^D / p_k^D - r_k^D. \tag{4.10}$$

The penalty function r_k^D is defined as follows:

$$r_k^D = \frac{R_k^D}{p_k^D} \exp\left(-\frac{p_k^D}{w_k^D}\right), \tag{4.11}$$

where w_k^D is the weight factor, which depends on three factors: 1) d_k's incoming data rate λ_k^D, 2) the total incoming data rate R_{in}^S, and 3) the maximum incoming data rate $R_{\max}^S$. As discussed earlier, a more weighted penalty leads to a higher power consumption for maximum utility. Therefore, the weight factor w_k^D increases based on the increase of two conditions.

- The ratio of the incoming data rate of DAP k to the total uplink date rate, that is, λ_k^D / R_{in}^S.
- The ratio of the total uplink data rate to the maximum uplink data rate, that is, $R_{in}^S / R_{\max}^S$.

The two conditions can be combined together as follows:

$$(\lambda_k^D / R_{in}^S)(R_{in}^S / R_{\max}^S) = \lambda_k^D / R_{\max}^S.$$

Without loss of generality, the weight factor for DGD k is defined as follows:

$$w_k^D = \frac{c}{p_{\max}^D} \left(\frac{\lambda_k^D}{(R_{\max}^S)(D + B)}\right)^2, \tag{4.12}$$

where c is a scaling factor so that the final results can be normalized into a reasonable range. Note that for a given λ_k^D, w_k^D is a constant.

In the previous section, the penalty function was defined to have three properties. We then check if the penalty function $r_k^D(\cdot)$ can meet all properties. Taking the first derivative with respect to p_k^D as follows:

$$\frac{\partial r_k^D}{\partial p_k^D} = -\frac{R_k^D[1 + p_k^D/w_k^D]e^{-p_k^D/w_k^D}}{(p_k^D)^2} < 0. \tag{4.13}$$

Therefore, $r_k^D(\cdot)$ is a monotonically decreasing function with respect to p_k^D. We then take the second derivative with respect to p_k^D as follows:

$$\frac{\partial^2(r_k^D)}{\partial(p_k^D)^2} = \frac{R_k^D e^{-p_k^D/w_k^D}[(p_k^D)^2 + 2p_k^D w_k^D + 2(w_k^D)^2]}{(p_k^D)^3(w_k^D)^2} > 0. \tag{4.14}$$

Therefore, $r_k^D(\cdot)$ is convex. And finally, the penalty function has the following property:

$$\lim_{p_k^D \to \infty} r_k^D = 0. \tag{4.15}$$

In summary, r_k^D in Eq. (4.11) satisfies all three properties required for a penalty function.

In the DGD level game, for an arbitrary and selfish player d_k, the goal is to find the optimal power to achieve its maximum power efficiency when counting in the penalty. That can be achieved by solving the following optimization problem:

$$\max_{p_k^D \in (0, p_{\max}^D]} u_k^D, \tag{4.16}$$

with given p_i^D for all $i \neq k$ and power consumption of all BGDs.

4.4.2 BGD Level Game

The BGD level game is defined as $\mathcal{G}_B = [\mathcal{B}, \{p_i^B\}, \{u_i^B(\cdot)\}]$. In this level of game, $\mathcal{B} = (b_1, \ldots, b_B)$ are the players (i.e. BGDs), $\{p_i^B\}$ is the action set (i.e. power consumption of each BGD), and $\{u_i^B(\cdot)\}$ is the set of individual utilities (i.e. power efficiency with defined penalty). Similarly, we assume that the maximum transmitting power $p_{\max}^B$ is identical to all the players. Given the DGDs power profile $\mathbf{p}_D = (p_1^D, p_2^D, \ldots, p_D^D)$ generated in the DGD level game, an arbitrary player b_k calculates its utility in the BGD level game as follows:

$$u_k^B = T_k^B/p_k^B - r_k^B, \tag{4.17}$$

where the penalty function is defined as follows:

$$r_k^B = \frac{R_k^B}{p_k^B} \exp\left(-\frac{p_k^B}{w_k^B}\right), \qquad (4.18)$$

where w_k^B is the weight factor of this penalty function calculated as

$$w_k^B = \frac{c}{p_{max}^B}\left(\frac{\lambda_k^B}{R_{max}^S/(D+B)}\right)^2. \qquad (4.19)$$

Similarly, this selfish player b_k aims to maximize its utility with respect to the penalty function by solving the following optimization problem:

$$\max_{p_k^B \in (0, p_{max}^B]} u_k^B, \qquad (4.20)$$

with given p_i^B for all $i \neq k$ and power consumption of all DGDs.

4.5 Analysis of the Proposed Schemes

In this section, we analyze the schemes proposed in the previous sections. In particular, we first study the numbers of DGDs and BGDs. Then we study the proposed two-level Stackelberg game.

4.5.1 Estimation of B and D

Before discussing the hierarchical uplink transmission power control scheme, we need to decide the actual number of players in each level of the Stackelberg game. Without loss of generality, when estimating D, we assume that the DGDs are identical to each other. With this assumption, the power consumption levels of DGDs are the same, that is, $p_i^D = p_D^D, h_i^D = h_D^D$. The uplink data rate is also identical at all DGDs, i.e. $R_i^D = R_0^D$, $\forall i \in [1, D]$. Note that these simplifications apply only to the estimations of D and B in this subsection. To clarify the illustration, we temporarily rewrite Eq. (3.28) as follows:

$$\tilde{\gamma}_k^D = \frac{p_k^D h_k^D}{\sigma^2 + \frac{D-1}{N}p_k^D h_k^D}. \qquad (4.21)$$

With a fixed noise σ^2 and taking the first derivative with respect to p_k^D as follows:

$$\frac{\partial \tilde{\gamma}_k^D}{\partial p_k^D} = \frac{\sigma^2 h_k^D}{[\sigma^2 + \frac{D-1}{N} p_k^D h_k^D]^2} \geq 0. \tag{4.22}$$

Therefore $\tilde{\gamma}_k^D$ is monotonically increasing with respect to p_k^D. This will help to estimate D and B. Recall that DGDs will transmit α portion of the total uplink data rate, as follows:

$$\sum_{i=1}^{D} R_i^D f_e(\gamma_{i\,\mathrm{max}}^D) = \alpha R_{\mathrm{max}}^S, \tag{4.23}$$

where R_i^D is the transmission rate of d_i and $\gamma_{i\,\mathrm{max}}^D$ is the maximum SINRs achieved by d_i at its maximum power p_{max}^D. Note that the maximum SINRs are also identical at each DGD:

$$\gamma_{i\,\mathrm{max}}^D = \tilde{\gamma}_{\mathrm{max}}^D, \ \forall \, i \in [1, D].$$

The maximum SINR $\tilde{\gamma}_{\mathrm{max}}^D$ is achieved when all DGDs transmit at the maximum power. Applying $\tilde{\gamma}_{\mathrm{max}}^D$ to Eq. (4.23) as follows:

$$DR_0^D f_e(\tilde{\gamma}_{\mathrm{max}}^D) = \alpha R_{\mathrm{max}}^S. \tag{4.24}$$

Eq. (4.24) may have two solutions, for example D_1 and D_2. Assuming $D_1 \leq D_2$, the number of DGD D is estimated as follows:

$$\tilde{D} = \lceil D_1 \rceil. \tag{4.25}$$

We then estimate B, the number of BGDs. Similarly, all the BGDs are assumed to be identical to each other. With the interference from DGDs added, the estimated SINR for b_k is calculated as follows:

$$\tilde{\gamma}_k^B = \frac{p_k^B h_k^B}{\sigma^2 + \frac{B-1}{N} p_k^B h_k^B + \frac{D}{N} p_i^D h_i^D}, \ \forall \, i \in [1, D]. \tag{4.26}$$

Without loss of generality, we further assume that the DGDs are also identical to BGDs in the estimation, where $p_i^D = p_j^B = p_B^B$ and $h_i^D = h_j^B = h_B^B$, $\forall i \in [1, D]$, $\forall j \in [1, B]$. Thus Eq. (4.26) can be represented as follows:

$$\tilde{\gamma}_k^B = \frac{p_k^B h_k^B}{\sigma^2 + \left(\frac{B-1}{N} + \frac{D}{N}\right) p_k^B h_k^B}. \tag{4.27}$$

Obviously, $\tilde{\gamma}_k^B$ is monotonically increasing with respect to p_k^B. Note that DGDs and BGDs together support all uplink transmissions:

$$\sum_{i=1}^{B} R_i^B f_e(\gamma_{i\max}^B) + \sum_{i=1}^{D} R_i^D f_e(\gamma_{i\max}^D) = (\alpha + \beta)R_{\max}^S. \quad (4.28)$$

If the transmission rates of BGDs and DGDs are identical to each other, where $R_i^B = R_j^D = R_0^B$, $\forall\, i \in [1, B]$, and $\forall\, j \in [1, D]$, the maximum SINRs are also identical, that is, $\gamma_{i\max}^B = \gamma_j^D = \tilde{\gamma}_{\max}^B$, $\forall\, i \in [1, B]$, and $\forall\, j \in [1, D]$. Eq. (4.28) can be represented as follows:

$$(B + D)R_0^B f_e(\gamma_{\max}^B) = (\alpha + \beta)R_{\max}^S. \quad (4.29)$$

Eq. (4.29) may have two solutions, B_1 and B_2. Assuming $B_1 \leq B_2$, the number of BGD B is estimated as follows:

$$\tilde{B} = \lceil B_1 \rceil. \quad (4.30)$$

For to make the illustration clearer, the estimation of DGDs and BGDs will be represented as $D = \tilde{D}$ and $B = \tilde{B}$ in the rest of the chapter.

4.5.2 Analysis of the Proposed Stackelberg Game

We apply the backward induction method to approach the proposed Stackelberg game. In a two-level Stackelberg game, since the follower's strategies will affect the leader's strategies, we first study the BGD level game $\mathcal{G}_B = [B, \{p_i^B\}, \{u_i^B(\cdot)\}]$. The goal is to find the Nash equilibrium (NE) for this game with a given action set of DGDs, that is, $\mathbf{p}_D = (p_1^D, p_2^D, \dots, p_D^D)$.

Definition 4.2 *Nash equilibrium (NE)* An action set $\mathbf{p}_B = (p_1^B, p_2^B, \dots, p_B^B)$ is an NE of game $\mathcal{G}_B = [B, \{p_i^B\}, \{u_i^B(\cdot)\}]$ if, $\forall\, i \in [0, B]$, $\forall p_j^B \in [0, p_{\max}^B]$, $u_i(p_i^B, \mathbf{p}_{-i}^B) \geq u_i(p_j^B, \mathbf{p}_{-i}^B)$, where $\mathbf{p}_{-i}^B = (p_1^B, \dots, p_{i-1}^B, \dots, p_{i+1}^B, \dots, p_B^B)$.

With a fixed noise, SINR γ is a function of transmit power p. Therefore, the efficiency function $f_e(\cdot)$ can be represented as a function of p. Given a weight factor, the penalty function $r(\cdot)$ is also a function of p. Moreover, all BGDs are assumed to have the same physical layer transmission rate, since they are equipped with the same hardware, that is, $R_i^B = R_0^B$, $\forall\, i \in [1, B]$. Without loss of generality, the transmission rate is normalized to 1. With a fixed noise n and weight factor w, the utility function Eq. (4.17) can be represented as follows:

$$u(p) = g(p) - h(p), \tag{4.31}$$

where $g(p) = f_e(p)/p = (1 - e^{-p/n})^M/p$ and $h(p) = r(p) = e^{-p/w}/p$.

Definition 4.3 *Quasiconcavity* *[76]* $\forall t \in R,\ \forall \alpha \in [0, 1],\ \forall x_1,$ $x_2 \in I \subset R, u(x_1) \geq t$ and $u(x_2) \geq t$ imply that

$$u(\alpha x_1 + (1 - \alpha)x_2) \geq t. \tag{4.32}$$

Lemma 4.1 Function $g(p)$ is quasiconcave.

Proof: Because function $f_e(p)$ is S-shaped, thus $g(p)$ is quasiconcave [76]. □

Lemma 4.2 Function $u(p)$ is quasiconcave.

Proof: Take the first derivative of $u(p)$, that is, $u'(p) = g'(p) - h'(p)$, we have two observations as follows:

$$\lim_{p \to 0} u'(p) = \lim_{p \to 0} \frac{(p + w)}{wp^2} = +\infty, \tag{4.33}$$

and

$$\lim_{p \to +\infty} u'(p) = \lim_{p \to +\infty} \left(-1 + \frac{M}{n}pe^{-p/n} + \frac{1}{w}pe^{-p/w} \right)/p^2$$
$$= \lim_{p \to +\infty} \frac{-1}{p^2}$$
$$= 0^-, \tag{4.34}$$

where 0^- indicates the negative approach to 0. As it shows, there is at least one p_z such that $u(p_z) = 0$. As shown in Figure 4.3, there is a unique zero point in function $u'(p)$. We then prove the uniqueness of this point.

According to Lemma 4.1, $g(p)$ is quasiconcave. Therefore, if $g'(p_g) = 0$ then $g'(p) > 0, \forall p \in (0, p_g)$; if $g'(p_g) < 0, \forall p \in (p_g, \infty)$, then there exists a point p_l where $g'(p)$ increases after $p = p_l$. Moreover, $g'(p_g)$ approaches 0 at infinity, as follows:

$$\lim_{p \to \infty} g'(p) = \lim_{p \to \infty} \left(-\frac{1}{p^2} + \frac{M}{n} \frac{e^{-p/n}}{p} \right) = 0^-. \tag{4.35}$$

Recall the properties of the penalty function that $h(p)$ is strictly decreasing and converges to 0. Therefore $h'(p) < 0$ and $h'(p)$ increasingly converges to 0^-, which indicates that $u'(g) > 0$, $\forall p \in (0, p_g)$. In addition, given $g'(p_z) < 0$ for all $p_z \in (p_l, \infty)$, we can see that,

$$\lim_{p \to \infty} \frac{h''(p)}{g''(p)} = 0, \tag{4.36}$$

which indicates that $h'(p)$ converges to 0 faster than $g'(p)$. Consequently, when $p > p_l$, we have $g'(p) - h'(p) < 0$, for all $p \in (p_l, \infty)$. If $p_z \geq p_l$, we have $u'(p) < 0$, for all $p \in (p_z, \infty)$. Clearly, $u'(p) > 0$ when $p < p_z$. This indicates that p_z is the unique solution to $u(p_z) = 0$. If $p_g < p_z \leq p_l$, it is clear that $u'(p) > 0$ for $p < p_g$. For $p_g < g \leq p_l$, since $g'(p)$ is decreasing and $h'(p)$ is increasing, thus $g'(p) < h'(p)$. For $p > p_l$, we always have $g'(p) < h'(p)$. Therefore, $u'(p) < 0$, $\forall p \in (p_z, \infty)$, and p_z is the only point such that $u'(p_z) = 0$. And $u(p)$ is strictly increasing for $p < p_z$, whereas $u(p)$ is strictly decreasing for $p > p_z$. Thus it proves the uniqueness of p_z such that $u(p_z) = 0$.

We next prove $u(p)$ is quasiconcave. Without loss of generality, assume $0 \leq p_1 < p_2$, and $t = \min\{u(p_1), u(p_2)\}$. Because $u(p)$ is continuous and strictly increasing in the interval $[0, p_z)$, there exists $\underline{p}_t$ such that $u(p) \geq t$, $\forall p \in (\underline{p}_t, p_z)$, and $u(p) < t$, $\forall p \in (0, \underline{p}_t)$. Similarly, since $u(p)$ is continuous and strictly decreasing in the interval (p_z, ∞), there exists $\bar{p}_t$ such that $u(p) < t$, $\forall p \in (\bar{p}_t, \infty)$. Then we have $u(p) \geq t$, $\forall p \in [\underline{p}_t, \bar{p}_t)$. Therefore, $t = \min\{u(p_1), u(p_2)\}$ indicates that $\{p_1, p_2\} \subset [\underline{p}_t, \bar{p}_t]$. Since $p_1 < \alpha p_1 + (1 - \alpha)p_2 < p_2$, $\forall \alpha \in (0, 1)$, we get $\underline{p}_t < \alpha p_1 + (1 - \alpha)p_2 < \bar{p}_t$, and $u(\alpha p_1 + (1 - \alpha)p_2) \geq t$. This completes the proof. $\qquad\square$

Let $f_r(p) = \exp(-p/w)$ and $f(p) = f_e(p) - f_r(p)$; the utility function Eq. (4.31) can be represented as follows:

$$u(p) = f(p)/p. \tag{4.37}$$

The utility function for BGD level game is $u_k^B = f(p_k^B)/p_k^B$ accordingly.

Corollary 4.1 The utility function $u_k^B(\cdot)$ has a unique maximizer $\tilde{p}_k^B$.

Proof: According to Lemma 4.2, $u_k^B(\cdot)$ is quasiconcave. Therefore, if we set $u_k'^B(\cdot) = [p_k^B f'(p_k^B) - f(p_k^B)]/(p_k^B)^2 = 0$, there is a unique solution $\tilde{p}_k^B$ to $f(p_k^B) = p_k^B f'(p_k^B)$, and it is the unique maximizer to u_k^B. $\qquad\square$

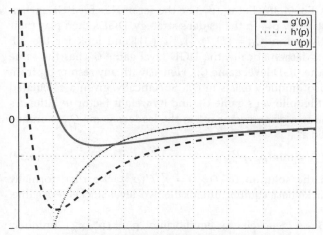

Figure 4.3 Illustration of $u'(p) = g'(p) - h'(p)$.

Corollary 4.2 Given $\mathbf{p}_D$ and weight factor w_i^B, b_i's best response to a given vector $\mathbf{p}_{-i}^B$ in the follower's game $\mathcal{G}_B$ is unique and calculated as

$$re_i^B(\mathbf{p}_{-i}^B) = \min(\tilde{p}_i^B, p_{\max}^B). \tag{4.38}$$

Proof: From Corollary 4.1, we know that $\tilde{p}_k^B$ is the unique maximizer to $u_k^B(\cdot)$. If $\tilde{p}_k^B \leq p_{\max}^B$, the best response is obviously $\tilde{p}_k^B$. If $\tilde{p}_k^B \geq p_{\max}^B$, the best response is $p_{\max}^B$ since u_k^B is strictly increasing when $p_k^B < \tilde{p}_k^B$. □

Theorem 4.1 A Nash equilibrium exists in noncooperative game $\mathcal{G}_B$.

Proof: Since u_k^B is quasiconcave in p_k^B and has a continuous derivation, hence an NE always exists [75]. □

Theorem 4.2 Noncooperative game $\mathcal{G}_B$ has a unique equilibrium.

Proof: According to Corollary 4.1, $\forall i \in [1, B]$, b_i has a unique best response p_k^{B*} with a given $\mathbf{p}_{-i}^B$. Moreover, an NE is guaranteed according to Theorem 4.1. Clearly, this NE is unique for $\mathcal{G}_B$. □

In summary, we know that the BGB level noncooperative game $\mathcal{G}_B = [\mathcal{B}, \{p_i^B\}, \{u_i^B(\cdot)\}]$ has a unique NE with a given $\mathbf{p}_D$, and it can be calculated by Eq. (4.38). Let $\mathbf{u}_B^*(\mathbf{p}_D)$ be the set of NEs for BGDs when DGDs play strategy $\mathbf{p}_D$. In the leader's strategy, DGDs then play their noncooperative game $\mathcal{G}_D = [D, \{p_i^D\}, \{u_i^D(\cdot)\}]$ and look for the NE based on $\mathbf{u}_B^*$. Mathematically, the DGD level game $\mathcal{G}_D$ has the same structure as the BGD level game $\mathcal{G}_B$. Hence, with any results $\mathbf{p}_B$ from $\mathcal{G}_B$, an NE also uniquely exists in $\mathcal{G}_D$. Specifically, given a calculated set $\mathbf{p}_B$ from the follower's game $\mathcal{G}_B$ and its weight factor w_i^D, the d_i's best response to a given vector $\mathbf{p}_{-i}^D$ in the leader's game $\mathcal{G}_D$ is unique and calculated as

$$re_i^D(\mathbf{p}_{-i}^D) = \min(\tilde{p}_i^D, p_{\max}^D), \tag{4.39}$$

where $\tilde{p}_i^D$ is the solution to $f(p_i^D) = p_i^D f'(p_i^D)$. We then formally define the Stackelberg equilibrium for the two-level Stackelberg game.

Theorem 4.3 A Stackelberg equilibrium $\mathbf{p}^* = (\mathbf{p}_D^*, \mathbf{p}_B^*)$ uniquely exists in the proposed two-level Stackelberg game.

Definition 4.4 *Stackelberg equilibrium (SE)* A vector $\mathbf{p}^* = (\mathbf{p}_D^*, \mathbf{p}_B^*)$ is an SE, if $\mathbf{p}_B^* \in \mathbf{u}_B^*(\mathbf{p}_D^*)$ and $\mathbf{p}_D^*$ is an NE for $\mathcal{G}_D = [D, \{p_i^D\}, \{u_i^D(\cdot)\}]$.

Proof: For a given $\mathbf{p}_D^*$, $\mathbf{p}_B^*$ exists and is unique. Since $\mathcal{G}_B$ and $\mathcal{G}_D$ are the same for analysis, for a given $\mathbf{p}_B^*$, $\mathbf{p}_D^*$ exists and is unique. $\quad\square$

In order to find the SE, Eq. (4.38) and Eq. (4.39) are computed reciprocally to find the NE for both games in each level as follows:

$$p_i^B = re_i^B(\mathbf{p}_{-i}^B, \mathbf{p}_D), \ \forall i \in [1, B],$$
$$p_j^D = re_j^D(\mathbf{p}_{-j}^D, \mathbf{p}_B), \ \forall j \in [1, D]. \tag{4.40}$$

4.5.3 Algorithms to Approach NE and SE

Based on the analysis above, we propose an algorithm to approach the NE in the two-level games. The BGD level game is presented in Algorightm 4.1. Switch $B \to D$ and $B \to D$, and the algorithm approaches the NE for the DGD level game $\mathcal{G}_D$.

To approach SE, we need to run Algorithm 4.1 reciprocally for both $\mathcal{G}_B$ and $\mathcal{G}_D$. This process is summarized in Algorithm 4.2.

Algorithm 4.1 Algorithm to approach NE for $\mathcal{G}_B$

Input: $\mathbf{p}_D$, λ_i^B, $\forall i \in [1,B]$, λ_i^D, $\forall i \in [1,D]$;
Output: $\mathbf{p}_B^*$;
1: Start with a random $\mathbf{p}_B$, where each $p_k^B \in (0, p_{\max}^B]$;
2: Calculate $re_i^B(\mathbf{p}_{-i}^B, \mathbf{p}_D)$ $\forall i \in [1,B]$;
3: **while** $\exists i \in [1,B]$, $re_i^B(\mathbf{p}_{-i}^D, \mathbf{p}_D) \neq p_i^B$ **do**
4: $\quad p_i^B \leftarrow re_i^B(\mathbf{p}_{-i}^B, \mathbf{p}_D)$, $\forall i \in [1,B]$;
5: $\quad$ Calculate $re_i^B(\mathbf{p}_{-i}^B, \mathbf{p}_D)$ $\forall i \in [1,B]$;
6: **end while**
7: $\mathbf{p}_B^* \leftarrow p_i^B$, $\forall i \in [1,B]$; // Output NE

Algorithm 4.2 Algorithm to approach SE

Input: $\mathbf{p}_D$, λ_i^B, $\forall i \in [1,B]$, λ_j^D, $\forall j \in [1,D]$;
Output: $\mathbf{p}^* = (\mathbf{p}_D^*, \mathbf{p}_B^*)$;
1: Calculate $\mathbf{p}_B^*$;
2: $\mathbf{p}_B \leftarrow \mathbf{p}_B^*$;
3: Calculate $\mathbf{p}_D^*$;
4: **while** $\exists i \in [1,D]$, $re_i^D(\mathbf{p}_{-i}^D, \mathbf{p}_B) \neq p_i^D$ **do**
5: $\quad p_i^D \leftarrow re_i^D(\mathbf{p}_{-i}^D, \mathbf{p}_B)$, $\forall i \in [1,D]$;
6: $\quad$ Calculate $\mathbf{p}_B^*$;
7: $\quad \mathbf{p}_B \leftarrow \mathbf{p}_B^*$;
8: $\quad$ Calculate $\mathbf{p}_D^*$;
9: **end while**
10: $\mathbf{p}_B^* \leftarrow p_i^B$, $\forall i \in [1,B]$; // Output SE
11: $\mathbf{p}_D^* \leftarrow p_i^D$, $\forall i \in [1,D]$; // Output SE

4.6 Numerical Results

4.6.1 Simulation Settings

In the simulation set up, the NAN is in an area of 1 km × 1 km, and it consists of 2000 smart meters, each having an adaptive sampling period 1-2 *sec*. Each sample generates 12 KByte, or 48–96 Kbps real time data rate and is finally sent to a DGD or BGD by following the HWMP routing protocol. Therefore, $R_{\max}^S = 192$ *Mbps* when all smart meters have the same sampling period at 1 sec. Noise $\sigma = 10^{-15}$ Watt (-120 dBm). The BGDs and DGDs have the same transmission rate at $R_0^D = R_0^B = 17$ Mbps according to the IEEE 802.16 protocol. A concentrator is located 1 kilometer away from the border of the NAN. The DGDs and BGDs are supposed to be deployed as close to the concentrator as possible in order to save transmitting energy. Without loss of generality, let the DGDs be 1 kilometer away from the concentrator, and BGDs be roughly 100 m further. Considering $A = 10^{-7}$, $v = 2$, the path gain of DGDs and BGDs are $h_i^D = 10^{-13}$, $\forall i \in [1, D]$ and $h_i^B = 8.26 \times 10^{-14}$, $\forall i \in [1, B]$ respectively. To calculate transmission efficiency, let processing gain $G = 128$ and packet length M=100 bits. Let $\alpha + \beta = 1$ so that the NAN can meet the maximum demand.

4.6.2 Estimate of *D* and *B*

With different values of α, the numbers of DGDs and BGDs have different estimates, as shown in Figure 4.4. Note that when $\alpha = 0$ or $\alpha = 1$, there is no hierarchical structure, and therefore the Stackelberg game approach is relaxed to a noncooperative game approach.

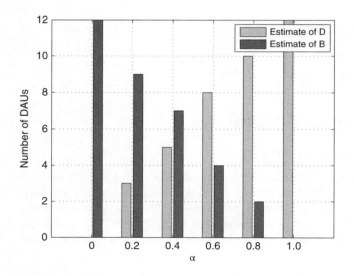

Figure 4.4 The estimate of *D* and *B* with respect to different α.

4.6.3 Data Rate Reliability Evaluation

Without loss of generality, assume $\alpha = 0.4$, which returns $D = 5$ and $B = 7$. We use the same Stackelberg approach to obtain the results targeting pure power efficiency utility (denoted as "no penalty" in simulation results). Figure 4.5 clearly shows that achieving pure power efficiency does not guarantee the reliability requirement.

We then discuss how the proposed utility performs with penalty. We first show the impact on the weight of the penalty function c. With a higher weight, the proposed scheme achieves a higher reliability when the smart meters generate more data. When $c = 0.25$ and $c = 0.3$, it always achieves 100% reliability. However, the question is which c to choose for deployment. Before answering this question, we need to show the power efficiency aspect of our proposed scheme. Specifically, we examine the impact of different incoming data rates on the total power usage with two sets of settings. One set is with $\alpha = 0$; in other words, the Stackelberg game is degraded to a noncooperative game only with $D = 0$, $B = 12$. (We keep the path gain settings such that 5 BGDs are closer to the concentrator.) The other set is with $\alpha = 0.4$, where $D = 5$ and $B = 7$, and the results are obtained by the Stackelberg game. Each point is the average value of 300 simulation runs.

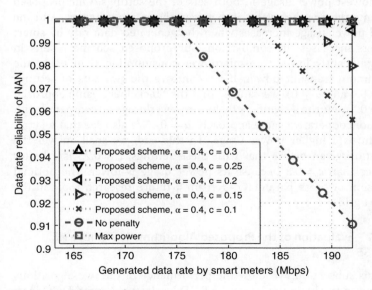

Figure 4.5 Reliability of NAN.

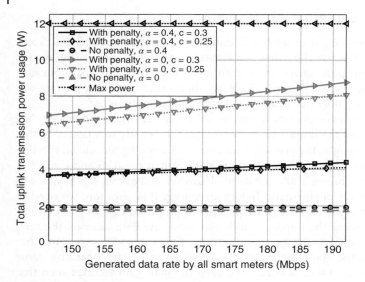

Figure 4.6 Total uplink transmission power usage comparison.

As shown in Figure 4.6, the pure power efficiency scheme returns the lowest power usage in both sets of the settings. Our proposed scheme returns the results in between those two schemes. And the total power usage increases when the generated data rate by smart meters increases. In comparison, our proposed scheme results in a significantly lower transmit power consumption than using the maximum transmit power. If we compare the two sets of settings ($\alpha = 0.4$ and $\alpha = 0$), we can see that the Stackelberg game approach ($\alpha = 0.4$) returns a significantly lower power usage as compared with the noncooperative game approach ($\alpha = 0$). We also see that a larger c returns a higher power usage. To this end, we shall answer the question given before that with these settings, $c = 0.25$ is a better choice than $c = 0.3$ because its power usage is lower. The optimal decision of those parameters remain a future research topic in the smart grid.

4.6.4 Evaluation of the Proposed Algorithms to Achieve NE and SE

In this subsection, we show the performance of the two algorithms proposed to obtain the NE in the BGD level game $\mathcal{G}_B$ and DGD level game $\mathcal{G}_D$ and to obtain the SE in the two-level Stackelberg game. The evaluation of the algorithms uses the same set of randomly generated incoming data rates applied to the previous analysis for consistency.

As shown in Figure 4.7, both BGD- and DGD-level noncooperative games can converge after very few iterations using Algorithm 4.1. As shown in Figure 4.8, the SE can converge after very few iterations using Algorithm 4.2.

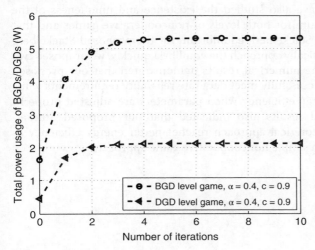

Figure 4.7 Convergence of the NE.

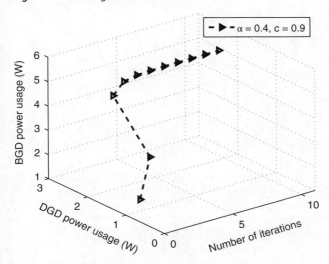

Figure 4.8 Convergence of the SE.

4.7 Summary

In this chapter, we proposed a hierarchical uplink transmission power control scheme with a penalty function using a two-level Stackelberg game-theoretical approach, in order to achieve both energy efficiency and data rate reliability requirements. For a linear receiver, we have also studied the existence and uniqueness of the Nash equilibrium for both levels of noncooperative games and that of the expand Stackelberg equilibrium for the proposed Stackelberg game. To practically approach the equilibria quickly, we proposed two algorithms. The numerical results demonstrated that our proposed scheme can successfully meet data rate reliability requirements while achieving energy efficiency when parameters are adjusted properly. The numerical results also indicated that our proposed Stackelberg game-theoretical approach reaches better energy efficiency as compared with simple noncooperative game approach.

5

Design and Analysis of a Wireless Monitoring Network for Transmission Lines in the Smart Grid

In this chapter, we design a monitoring network for the transmission lines in the smart grid, taking advantage of the existing utility backbone network. We propose several centralized power allocation schemes to set the benchmark for deployment and operation of the wireless sensor nodes in the monitoring system. A distributed power allocation scheme is proposed to further improve the power efficiency of each sensor node for realistic cases with dynamic traffic. Numerical results and a case study are provided to demonstrate the proposed schemes.

5.1 Background and Related Work

5.1.1 Background and Motivation

The power transmission line is an important part of the power grid. It is critical for the control center to monitor the status of the transmission line (e.g. the working status of the key components, the voltage and phase of the power, etc.), and its surrounding environment (e.g. the temperature, the humidity, etc) so that necessary actions can be taken in time to prevent disastrous outcomes. However, monitoring the transmission line is not an easy task for a traditional power grid, because of its vast expanse as well as because parts of it are deployed in remote areas, such as deserts and mountains.

The smart grid, which integrates an advanced communication system to the traditional power grid [4, 6–8, 10, 34, 82–84], makes it possible to install and upgrade the monitoring equipment along transmission lines to obtain precise and timely status information.

Smart Grid Communication Infrastructures: Big Data, Cloud Computing, and Security,
First Edition. Feng Ye, Yi Qian, and Rose Qingyang Hu.
© 2018 John Wiley & Sons Ltd. Published 2018 by John Wiley & Sons Ltd.

In the construction of power transmission lines, an optical ground wire (OPGW) [85] is usually installed alongside them [86]. Therefore, the monitoring data can be transmitted to the control center almost immediately once the data gets access to OPGW. However, it is not economical to deploy an OPGW gateway at each transmission tower. In practice, the gateways are deployed far away from each other (e.g. every 20 kilometers) [87].

In this chapter, we design a monitoring network for the transmissionline, based on the existing OPGW. The designed monitoring network consists of hundreds of *wireless sensor nodes* that are deployed on top of selected power line transmission towers. Each wireless sensor is a piece of comprehensive hardware which includes several physical sensors and a wireless transmission module. The physical sensors (e.g. current transformers [88, 89]) are deployed to measure the electric current status (e.g. electric flow, phase status, etc.) and line positions (e.g. sag) [90–92]. The wireless transmission module delivers the monitored data to the control center for further analysis. Wireless technology has been widely adopted in the smart grid. For example, the smart grid in China has been allocated a dedicated spectrum for wireless transmission [93, 94].

Although the transmission line carries electricity, it is too powerful to directly support a wireless sensor node. Therefore, we assume that the wireless sensors are powered by batteries coupled with green energy generation (e.g. solar power or wind turbines). Green energy and energy harvesting techniques have been studied to show their availability to support general wireless sensor networks as well as communication networks in the smart grid [10, 95, 96]. With green energy, the wireless monitoring network can be deployed in a more flexible way. Now that the sensors are powered by green energy, we need to design a proper power supply component (the size of solar panel and the capacity of the battery) for each sensor node so that the deployment and maintenance costs can be budgeted without much waste. Note that the physical sensors are assumed to be powered at a constant rate thus our focus is on the wireless transmission module.

The most imperative requirement of the transmission line monitoring network is timely delivery of data while being energy efficient. To tackle this issue, we propose several centralized schemes to estimate the power consumption of each sensor. Specifically, the first scheme is designed to minimize the total power consumption of all sensors. Due to its complexity, we relax the approach to minimizing the total transmission efficiency while meeting the latency requirement. We also introduce a game theoretical scheme for faster computation and two more schemes with partial fixed variables for practical application to a large network. The results of the centralized schemes are used as benchmarks for network deployment and operational settings. In practical field operations, the gathered/generated monitoring data is not always at its maximum for each sensor, and the frequency of gathering such data is lower than that of the worst-case scenario considered in the centralized schemes. Moreover, centralized schemes are not applicable in practical operations due to the real-time transmission of the monitoring network. Therefore, we propose an adaptive distributed power allocation scheme so that real-time transmission can be achieved while each sensor can be more energy efficient by taking into consideration the dynamic transmitting data. The proposed distributed scheme is guaranteed to be more power efficient since it is bounded by the benchmark settings from the centralized schemes.

5.1.2 Related Work

The cyber-physical system of the smart grid, which has been studied by many researchers [34, 6, 8, 10, 97] includes data transmission, power management, cybersecurity, etc. Only a few researchers works mention the monitoring network of the power line transmission system. In most cases, the monitoring systems for high-voltage power transmission lines takes into consideration both electric current and the positions of the transmission lines so that overload, phase unbalance, fluctuation, etc. can be avoided or reduced, and some situations such as sagging and galloping can be tracked by the control center.

Current transformers (CTs) are typically used for current measurement [88, 89]. Some existing devices can directly or indirectly measure the sag of transmission lines. For example, Ren et al. proposed a dynamic line rating system to monitor the line sag under complex climates (e.g. heavy rains, heavy snow, strong wind, etc.) [90]. Huang et al. proposed an on-line scheme to monitor the icing density and type of the transmission line [92]. Sun et al. proposed a high-voltage transmission-line monitoring system based on magnetoresistive sensors [91] that calculate both the current flow and the line positions from the magnetic field of the phase conductors. Although those monitoring systems work properly for a section of transmission lines, they lack a data transmission technology to deliver the measured data to the control center over long distances in a short period of time. Therefore, the control systems of the smart grid, such as updated supervisory control and data acquisition (SCADA) and general network operations center (NOC), are not fully utilized to provide real-time monitoring and reaction.

Instead of monitoring technologies for a specific component or a section between two towers, our focus is on the long-distance data delivery of the monitoring network's information to the control center so that the advance control system in the smart grid can take action quickly when necessary. Lin et al. proposed a power transmission line monitoring frame based on a wireless sensor netword [97]. A traditional wireless mesh sensor network was studied due to multiple sensors on each tower. Routing protocols and hybrid medium access control (MAC) were also proposed to increase the energy efficiency of the sensor network. However, a renewable energy source was not considered, and the transmission data rate was not guaranteed by the proposed schemes. Energy efficiency in wireless sensor networks and multiaccess networks has also been widely studied [74–76]. However, traditional approaches generally assume that the data is delay tolerant and the nodes are subject to unrecoverable failures (e.g. a fully discharged battery). Therefore, we cannot apply the power allocation schemes directly from those works. In our work, we propose a utility function that not only targets the energy efficiency of the wireless transmission module but is also adaptive to different link delay and data traffic requirements.

The rest of this chapter is organized as follows. In Section 5.2, we present the transmission line monitoring network model in the future smart grid. In Section 5.3, we propose several centralized power allocation schemes in which the results can be used as a benchmark for the sensor network design. We also propose a distributed scheme based on the benchmark to deal with dynamic traffic loads for practical cases. In Section 5.6, we show the analysis and numerical results for the centralized schemes. We also conduct a case study to demonstrate the performance of the distributed scheme. We conclude the work in Section 5.7.

5.2 Network Model

In the smart grid, an OPGW network is deployed in parallel with a power transmission line and connected to the control center [85, 86].

A gateway to the OPGW network is deployed every few miles so that uploading and downloading of data to and from the data center can be achieved even if the power line towers are in remote areas. The proposed transmission line monitoring network consists of hundreds of *wireless sensor nodes* (we will use both *sensor* and *transmitter* interchangeably for simplicity hereafter) that are deployed on top of selected towers. The wireless sensors are powered by green energy (e.g. solar power or wind turbine); thus a wireless monitoring network can be deployed in a more flexible way. The combination of wireless technology and green energy also makes it more convenient to maintain or upgrade the sensors if it becomes necessary in the future. In the monitoring system, each sensor gathers the operating status of the transmission line and monitors its surrounding environment. Each sensor also delivers its monitoring data to the control center through the nearest OPGW gateway (gateway hereafter). Due to the long distance between two neighboring gateways, the monitoring network in between forms a multihop wireless network with a physical chain topology. An OPGW gateway usually has its own power supply from a reliable power source, and it is physically attached to the OPGW. In this chapter, our focus is on the multihop wireless part of the monitoring network between two neighboring gateways.

Figure 5.1 shows the studied section of the wireless monitoring network. As shown in the figure, r_1 and r_2 are two neighboring OPGW gateways. The transmission between r_1 and r_2 is achieved by two multihop wireless networks consisting of sensors. It is not necessary to mount sensors on all the towers; for simplicity, we show only the towers with sensors in Figure 5.1. Let $\mathcal{T} = \{\mathcal{T}_1, \mathcal{T}_2\}$ be the set of total sensors between r_1 and r_2, where $\mathcal{T}_1 = \{t_1, t_2, \ldots, t_T\}$ includes $T = |\mathcal{T}_1|$ sensors that that form a network uploading data to gateway r_1, and $\mathcal{T}_2 = \{t_{T+1}, \ldots, t_{2T}\}$ includes another T sensors that form a network uploading data to gateway r_2. Without loss of generality, we focus on

Control center

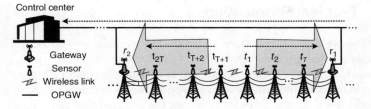

Figure 5.1 A section between two towers with fiber-optic connections.

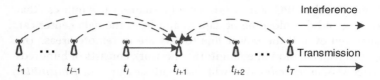

Figure 5.2 Source of interference for link $t_i \rightarrow t_{i+1}$

the sensors in $\mathcal{T}_1$ in the rest of the discussion. Note that the distance between two towers should be based on practical deployment. As mentioned before, the sensors upload their monitoring data in a multihop way; that is, sensor t_i uploads its data to its neighbor t_{i+1}, and t_{i+1} aggregates its own data to the data received from t_i and then sends the aggregated data to t_{i+2}. All the data from $\mathcal{T}_1$ will be uploaded to r_1 and then be transmitted to the control center through OPGW. Since the transmission over OPGW is much faster than the wireless proportion and is properly powered and well maintained, we thus focus on the power allocation over the sensors of the wireless multihop network. For simplicity, we assume that the networks work over half duplex transmission. Moreover, the focus of the monitoring network is on its uplink transmission, during which the sensors deliver data to the gateway.

For the transmission link $t_i \rightarrow t_{i+1}$, it is similar to the multiaccess wireless system at receiver t_{i+1}. The sensors in the monitoring network keep sending data; therefore t_{i+1} not only receives the signal from t_i but also the interference from other sensors, as shown in Figure 5.2. In general, t_{i+1} receives interference not only from $\mathcal{T}_1$ but also from other sensors, such as those in $\mathcal{T}_2$ and beyond. Since the sensors are placed apart from each other, interference from those sensors several hops away is low in practice. For better illustration, we only consider the interference from the sensors in $\mathcal{T}_1$ in this chapter.

5.3 Problem Formulation

To make the illustration clearer, we list the key variables and notations in Table 5.1 that are used throughout the rest of the chapter.

In wireless transmissions, additional energy consumption changes the fundamental tradeoff between energy efficiency and data rate [81]. The effective data rate takes into account the transmission error, data retransmission, packet loss, etc. Generally, with a given situation,

Table 5.1 Key sets and variables.

Sets	
$\mathcal{T}$	set of sensors/transmitters
p	set of transmitting power of all $t_i \in \mathcal{T}$
h	set of path gains
Variables	
l_i	gathered and generated data at t_i
d_i	outgoing data from t_i to t_{i+1}
p_i	transmitting power of t_i
γ_i	output SINR measured at t_{i+1} for t_i
$h_{i,j}$	path gain of t_i at t_j

a transmitter is required to transmit at a higher power in order to achieve a higher signal-to-interference-plus-noise ratio (SINR). And a higher SINR leads to a higher transmission rate, since the probability of successful transmission increases with respect to SINR. For a transmitting sensor t_i, its effective transmission rate at the receiving sensor t_{i+1} is widely adopted as [74–76]

$$T_i = R_i f(\gamma_i), \tag{5.1}$$

where R_i is the theoretical transmission rate, and γ_i is the SINR of t_i at t_{i+1}, which is computed as

$$\gamma_i = \frac{p_i h_{i,i+1}}{\sigma^2 + \frac{1}{N} \sum_{j \neq i, i+1} p_j h_{i,j}}. \tag{5.2}$$

In Eq. (5.1), $f(\gamma_i)$ is the efficiency function (or normalized transmission data rate), which is assumed to be increasing, continuous, and S-shaped (sigmoidal; more specifically, there is a point above which the function is concave and below which the function is convex [76]) with $f(+\infty) = 1$ and $f(0) = 0$. This efficiency function is commonly adopted as [74–76]

$$f(\gamma_i) = (1 - BER_i)^M, \; \gamma_i \geq 0, \tag{5.3}$$

where M is the packet length. Without loss of generality, we take FSK as an example for discussion. Then we have

$$BER_i = e^{-\gamma_i}. \tag{5.4}$$

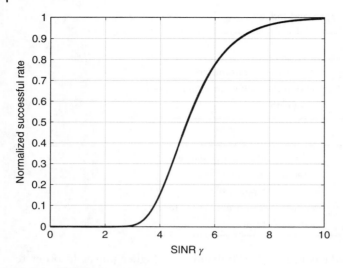

Figure 5.3 Illustration of $f(\gamma)$.

Figure 5.3 shows the S-shaped curve of the efficiency function Eq. (5.3). In Eq. (5.2), p_i is the transmission power of t_i, where

$$p_i \leq p_{\max}, \tag{5.5}$$

and $p_{\max}$ is the maximum transmission power. σ in Eq. (5.2) is the variance of the zero-mean white Gaussian noise, $h_{i,j}$ is the path gain of t_i at t_j, and N is the processing gain. Note that t_{i+1} does not cause interference to t_i due to the assumption of half-duplex transmission. Moreover, it is practical to assume that the transmitter and receiver pair have a line-of-sight (LOS) link. Then the path gain of a transmission link is mainly based on carrier frequency and physical distance between the transmitter and the receiver. Therefore, for simplicity, we adopt the free space path loss model to calculate $h_{i,j}$ as

$$h_{i,j}(dB) = 20 \log_{10}(d) + 20 \log_{10}(f) + 32.45, \tag{5.6}$$

where d in kilometers is the distance between two transmitters. Note that the distance between two transmitters may not be the distance between two consecutive towers, since sensors are not required for every tower. f in MHz is the transmission frequency. In the monitoring network, the traffic flow in each link is discrete instead continuous, since the transmission data rate is assumed to be much higher than the rate of data gathering and generation (gathering hereafter). Let f_i be the maximum frequency for t_i to transmit data to t_{i+1}; such data includes the data gathered by t_i and the data relayed from the previous transmitter t_{i-1}. Note that f_i is generally higher than the data-gathering frequency for t_i, since the data relayed from t_{i-1} is unpredictable. Let l_i be the length of data gathered each time by following frequency f_i. Since each transmitter aggregates the data from its predecessor, the total data to be transmitted from t_i to t_{i+1} is calculated as follows:

$$d_i = \sum_{k=1}^{i} l_k. \tag{5.7}$$

Combined with Eq. (5.1), the transmission delay from t_i to t_{i+1} is calculated as

$$\tau_i = \frac{d_i}{T_i}. \tag{5.8}$$

Each sensor t_i must finish the transmission of its aggregated data to t_{i+1} before it transmits new data; therefore τ_i is bounded as follows:

$$\tau_i \leq \frac{1}{f_i}. \tag{5.9}$$

Let $\bar{\tau}$ be the end-to-end delay for the data gathered by t_i to be successfully delivered to r_1 when each transmitter t_i always appends the aggregated data from t_{i-1} to the end of its own gathered data l_i. In this case, l_1 will suffer the longest end-to-end delay among all the data. Obviously, within $\bar{\tau}$, all data generated in a time period with respect to f_i for all t_i will be delivered to r_1. Let $\tau = \sum_{i=1}^{T} \tau_i$; in order to guarantee the successful delivery of all data in a time period, the system must meet the following requirement:

$$\tau \leq \bar{\tau}. \tag{5.10}$$

5.4 Proposed Power Allocation Schemes

In general, the monitoring data must be delivered to the control center within a short period of time. Therefore, the control center is not capable of gathering the real-time status (e.g. l_i) of each sensor and computing a centralized optimal power allocation accordingly. Nonetheless, a centralized scheme can be feasible with the scenario described below. With an optimal power allocation predetermined, it is helpful to reduce the cost when deploying the sensors. Especially for sensors powered by green energy, solutions generated from the centralized schemes can be used to design the power supply equipment (e.g. solar panel size and battery capacity). In the rest of this section, we first propose several centralized power allocation schemes while taking into consideration the worst-case scenario described below.

Definition 5.1 *Worst-case scenario*   In this case, all the sensors gather and send data at their maximum frequency f_i, and the data gathered from a transmitter is also at its maximum length, where $l_i = L_i$ for all $t_i \in \mathcal{T}$.

5.4.1   Minimizing Total Power Usage

Since the sensor nodes in the multihop wireless portion of the monitoring network are powered by green energy, it is important to reduce the power consumption of the sensors to lower the hardware cost. Therefore, the first centralized scheme $\mathcal{P}1$ is designed to minimize the total power consumption of the sensors that upload data to the same gateway (e.g. $\forall t_i \in \mathcal{T}_1$). The scheme is shown below.

$$\mathcal{P}1 : \ \min \sum_{i=1}^{T} p_i, \tag{5.11}$$

subject to Eq. (5.5), Eq. (5.9), and Eq. (5.10), $\forall t_i \in \mathcal{T}_1$.

However, there is no polynomial time algorithm to solve $\mathcal{P}1$. We then relax some of the constraints. As indicated before, the end-to-end delay requirement $\overline{\tau}$ is very small (e.g. 10 ms), however $1/f_i$ is relatively large (e.g. 1 s), so the constraint in Eq. (5.9) is implicitly satisfied in general. For the end-to-end delay, intuitively, if τ gets higher, the solution to $\mathcal{P}1$ will be small, since each sensor is able to use less power to transmit and achieve the same delay requirement. Therefore, we relax the constraint shown in Eq. (5.10) to the following equation

$$\tau = \sum_{i=1}^{T} \tau_i = \sum_{i=1}^{T} \frac{d_i}{T_i} = \sum \frac{d_i}{R_i} \left(\frac{1}{f(\gamma_i)} \right) = \overline{\tau}. \tag{5.12}$$

Note that $\frac{d_i}{R_i}$ is a constant in $\mathcal{P}1$ in the worst-case scenario. Therefore, Eq. (5.12) is an equation with variables $f(\gamma_i)$. As discussed before, $f(\gamma_i)$ is an increasing function with respect to

γ_i, and $\gamma_i(p_i)$ is also an increasing function with respect to p_i if $\mathbf{p}_{-i} = \{p_1, p_2, \ldots, p_{i-1}, p_{i+1}, \ldots, p_T\}$ is given. Therefore, minimizing the total transmission power can be viewed as minimizing the summation of the normalized transmission data rate of each link. We write the relaxed problem into $\mathcal{P}1'$ as shown below with constraints.

$$\mathcal{P}1' : \quad \min \sum_{i=1}^{T} f(\gamma_i), \tag{5.13}$$

subject to Eq. (5.12) and

$$f(\gamma_i) \geq \frac{d_i}{R_i} f_i, \tag{5.14}$$

$$f(\gamma_i) \leq f(\gamma_i^{\max}). \tag{5.15}$$

Constraint Eq. (5.14) is rewritten from Eq. (5.9) based on Eq. (5.1). It shows the lower bound of the normalized transmission rate of each link, below which the link would fail to deliver data before the transmission of new incoming data. Constraint Eq. (5.15) shows the upper bound of the normalized transmission rate of each link. The maximum SINR $\gamma_i^{\max}$ is achieved when $p_i = p_{\max}$. Since the objective is to minimize the summation of $f(\gamma_i)$, we assume $\gamma_i^{\max}$ is estimated when $p_k = p_{\max}$ for all $t_k \in \mathcal{T}_1$. Now that $\mathcal{P}1'$ is simpler compared with $\mathcal{P}1$, it is faster to find the optimal solution. The numerical results in the later section will demonstrate that. However, there is still no polynomial algorithm to achieve the optimal solution to $\mathcal{P}1'$. Therefore, if the network size is large, then it will take much more computational time to get the solution.

5.4.2 Maximizing Power Efficiency

In order to quickly converge to a solution, we propose a game theoretical scheme. A noncooperative game is formed as $\mathcal{G} = [\mathcal{T}_1, \{p_i\}, \{u_i(\cdot)\}]$, where $\mathcal{T}_1$ is the set of sensors or players, $\{p_i\}$ is the strategy set of each player, and $\{u_i(\cdot)\}$ is the corresponding utility. In traditional multiaccess wireless networks, the utility function is generally defined as the effective transmission rate over transmission power. However, such modeling does not work for the proposed monitoring network for two reasons. One is that the proposed monitoring network has a much more stringent delay requirement, while the traditional multiaccess wireless network is assumed to be more delay tolerant. The other reason is that the link requirement (e.g. SINR) cannot be simply assumed as uniform or fixed if an optimal (or suboptimal) solution is the goal.

In our proposed game for the monitoring network, two additional requirements must be satisfied for the utility function compared with the traditional one. One is that the maximizer of the utility function must be delay sensitive. Specifically, when τ_i is lower, the maximizer γ_i^* should be larger so that the effective transmission rate will be higher to adapt to the lower delay. The other requirement of the utility function is that the maximizer must respond to the dynamic traffic flow. That is to say, when d_i is higher, the maximizer γ_i^* should be larger for a higher effective transmission rate. Therefore, we update the utility function for sensor t_i into the one shown below,

$$u_i = \frac{(T_i)^{k_i d_i / \tau_i^*}}{p_i}, \tag{5.16}$$

where k_i is a weight factor and $\tau_i^\star$ is the benchmark link delay, which can be the optimal solution to $\mathcal{P}1$ or $\mathcal{P}1'$.

For each sensor t_i, the goal is to maximize the utility, as shown in $\mathcal{P}2$.

$\mathcal{P}2$: max u_i,

subject to Eq. (5.5), Eq. (5.9), $\tag{5.17}$

$$\tau_i \leq \tau_i^\star, \tag{5.18}$$

Lemma 5.1 Eq. (5.16) is quasiconcave with respect to γ.

Proof: Let $g(\gamma) = (T_i)^{k_i d_i / \tau_i}$; then

$$g(\gamma) = R^{k_i d_i / \tau_i} f(1 - e^{-\gamma})^{M k_i d_i / \tau_i}. \tag{5.19}$$

Since $f(\gamma)$ is S-shaped, and R, k_i, d_i, τ_i, M are determined values, then $g(\gamma)$ is also S-shaped. According to Eq. (5.2), p_i equals γ_i multiplied by a scalar. Therefore Eq. (5.16) is quasiconcave with respect to γ_i [76]. □

Corollary 5.1 The utility function $u_i(\cdot)$ has a unique maximizer γ_i^* such that $u_i(\gamma_i^\star) \geq u_i(\gamma_i)$, $\gamma_i \leq \gamma_i^{\max}$.

Proof: According to Lemma 5.1, $u_i(\cdot)$ is quasiconcave. Therefore, if we set $u_i'(\cdot) = [\gamma_i g'(\gamma_i) - g(\gamma_i)]/(\gamma_i)^2 = 0$, there is a unique solution γ_i^* to $g(\gamma_i) = \gamma_i g'(\gamma_i)$, and it is the unique maximizer to $u_i(\cdot)$. □

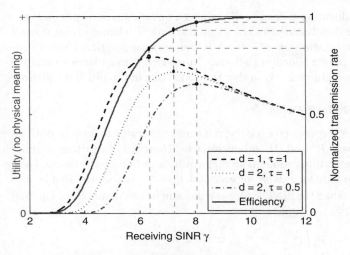

Figure 5.4 Illustration of the modified utility function.

Figure 5.4 illustrates the unique maximizer γ_i^* and the impact of dynamic d_i and benchmark $\tau_i^\star$ on the maximizer. Obviously, when d_i gets higher or $\tau_i^\star$ gets lower, the maximizer γ_i^* (if one uniquely exists) gets larger, and the corresponding effective transmission rate gets higher. Note that the unit of the utility function is not bit/joule.

Definition 5.2 *Nash equilibrium (NE)* A power vector $\mathbf{p}^* = (p_1^*, p_2^*, \ldots, p_T^*)$ is an NE of $\mathcal{G} = [\mathcal{T}_1, \{p_i\}, \{u_i(\cdot)\}]$ if, $\forall i \in [1, T]$, $\forall p_j \in (0, p_{\max}]$, $u_i(p_i^*, \mathbf{p}_{-i}^\star) \geq u_i(p_j, \mathbf{p}_{-i}^\star)$, where $\mathbf{p}_{-i}^\star = (p_1^*, \ldots, p_{i-1}^*, \ldots, p_{i+1}^*, \ldots, p_T^*)$.

Theorem 5.1 An NE $\mathbf{p}^* = (p_1^*, p_2^*, \ldots, p_T^*)$ uniquely exists in the proposed noncooperative game $\mathcal{G}$.

Proof: Because $u_i(\gamma)$ is continuously differentiable and quasiconcave over $\gamma_i > 0$, an NE always exists. Moreover, according to Corollary 5.1, with a given $\mathbf{p}_{-i}$ $\forall t_i \in \mathcal{T}_1$, t_i has a unique best response p_i^* so that $u_i(p_i^*, \mathbf{p}) \geq u_i(p_j, \mathbf{p})$, $\forall p_j \in (0, p_{\max}]$. Hence this NE is unique to $\mathcal{G}$. □

Compared with $\mathcal{P}1$ and $\mathcal{P}1'$, the advantage of the proposed game theoretical scheme is that it is capable of finding the solution (the NE) quickly by finding γ_i^* for all t_i and then computing the corresponding power allocation p_i for all t_i by solving a linear equation system. However, the disadvantage of the game theoretical scheme is that it must have the benchmark τ_i^*, which is the optimal solution from a centralized power allocation scheme. Moreover, it must have a carefully chosen weight factor k_i so that the constraint in Eq. (5.18) is satisfied.

5.4.3 Uniform Delay

When the network size is large, it is time consuming to find the optimal solutions to $\mathcal{P}1$ and $\mathcal{P}1'$ as benchmarks. Therefore, we further simplify the optimization problems so that a quick solution can be found. First, we assume that the delay for each link $t_i \to t_{i+1}$ is uniform, that is, $\tau_i = \tau_0 = \frac{\bar{\tau}}{|\mathcal{T}|}$. Then the SINR γ_i for each link can be calculated from Eq. (5.3) and Eq. (5.8) as follows:

$$\gamma_i = \underset{\gamma > 0}{\arg} \left((1 - e^{-\gamma})^M = \frac{\sum_{i \in \mathcal{T}} L_i}{\tau_i} \right). \tag{5.20}$$

Moreover, Eq. (5.2) can be represented as follows:

$$p_i h_{i,i+1} - \frac{\gamma_i}{N} \sum_{j \neq i, i+1} p_j h_{i,j} - \gamma_i \sigma^2 = 0. \tag{5.21}$$

Therefore, with a given γ_i for all t_i, the set of power allocation **p** can be easily calculated by solving a linear equation system where the number of equations and the number of variables are both T. Note that the last link is required to have an effective transmission rate of $T_{|\mathcal{T}|} = \frac{\sum_{i=1}^{T} d_i}{d_1} T_1$. Compared with the first link, $T_{|\mathcal{T}|}$ could be several times higher than T_1. However, since the packet from each sensor is small, and T_1 is possibly several Kbps only, it is still reasonable for the monitoring network to apply equal delay in each transmission link.

5.4.4 Uniform Transmission Rate

If the data generated by a sensor node increases to some extent, a fixed link delay may no longer be a reasonable assumption. Alternatively, we can fix the transmission rate of each link, that is for all $t_i \in \mathcal{T}$, as follows:

$$T_i = T_1,$$
$$\Rightarrow \frac{d_i}{\tau_i} = \frac{d_1}{\tau_1},$$
$$\Rightarrow \tau_i = \frac{L_1}{\tau_1 d_i}. \tag{5.22}$$

Without loss of generality, assume that $L_i = \max\{l_i\} = L_1$, then $\tau_i = i \times \tau_1$. From Eq. (5.10) we can get

$$\sum_{i \in \mathcal{T}} \tau_i \le \bar{\tau}$$

$$\Rightarrow \frac{|\mathcal{T}|(|\mathcal{T}| + 1)}{2} \tau_1 \le \bar{\tau}$$

$$\Rightarrow \tau_1 \le \frac{2\bar{\tau}}{|\mathcal{T}|(|\mathcal{T}| + 1)} \tag{5.23}$$

As stated before, Eq. (5.23) can be relaxed into equality, so that $\tau_1 = \frac{2\bar{\tau}}{|\mathcal{T}|(|\mathcal{T}|+1)}$. With τ_i and d_i given, T_i can be easily computed and also γ_i. Then the corresponding power allocation $\mathbf{p}$ can be found by solving the linear equation system Eq. (5.21).

5.5 Distributed Power Allocation Schemes

In field operations, centralized schemes are not applicable due to real-time transmission of the monitoring network. Moreover, each sensor gathers data less frequently, and the data size is usually smaller than its maximum value in the worst-case scenario. Therefore, although the radios must be active almost all the time, it is possible to have a distributed power allocation scheme based on dynamic data traffic to further reduce the power usage while allowing real-time transmission. For each sensor t_i, the most energy-efficient power allocation is to use the minimum transmission power while delivering the data on time, that is,

$$\mathcal{P}3 : \quad \min p_i \tag{5.24}$$

subject to Eq. (5.5), Eq. (5.9), and Eq. (5.18).

However, without full knowledge of the network, it is impractical for t_i to achieve an optimal solution. Moreover, the allocation of power must occur in a short time. Therefore, the goal for solving $\mathcal{P}3$ is to save as much power as possible while meeting the delay requirement at t_i.

Assume that the optimal power allocation $p_i^\star$ and delay requirement $\tau_i^\star$ are predetermined based on the worst-case scenario before the deployment of the network (e.g. the solution to $\mathcal{P}1$, $\mathcal{P}1'$, or $\mathcal{P}2$). Note that the proposed distributed scheme is based on the dynamic data traffic. Therefore, sensor t_i stores the length of the data previously sent as d_i' for comparison. Before t_i transmits data d_i, t_i checks its current

d_i. t_i first determines $T_i = \frac{d_i}{\tau_i^*}$. Then if $d_i > d_i'$, before t_i transmits the data, it sends a SINR request (a "HELLO" message) to t_{i+1} at power $p_i = p_i^*$. At the same time, t_{i+1} senses γ_i^* at its current status and replies with it to t_i. With given γ_i^*, t_i calculates the noise plus interference at t_{i+1} as

$$I_i = \left(\sigma^2 + \sum_{j \neq i, i+1} p_j h_{i,j} \right) = \frac{p_i h_{i,i+1}}{\gamma_i^*}. \tag{5.25}$$

Then t_i calculates adequate SINR as

$$\gamma_i = \arg_{\gamma > 0} (R_i(1 - e^{-\gamma})^M = T_i). \tag{5.26}$$

Obviously, $\gamma_i \leq \gamma_i^*$. Then t_i determines its current transmission power as

$$p_i = \arg_{p \leq p_i^*} \left(\frac{p_i h_{i,i+1}}{\sigma^2 + \sum_{j \neq i, i+1} p_i h_{i,j}} = \gamma_i \right). \tag{5.27}$$

If $d_i \leq d_i'$, then t_i keeps transmitting at current p_i and follows the process stated above to update its p_i accordingly. The scheme is summarized in Algorithm 5.1.

Algorithm 5.1 Dynamic transmitting power allocation

Input: p_i^*, d_i;
Output: p_i;
1: $T_i \leftarrow \frac{d_i}{\tau_i}$
2: **if** $d_i > d_i'$ **then**
3: $\quad p_i \leftarrow p_i^*$
4: **else**
5: $\quad$ Keep current p_i.
6: **end if**
7: t_i sends a "HELLO" message at p_i
8: t_{i+1} senses γ_i^* and sends it to t_i
9: t_i calculates I_i using Eq. (5.25)
10: t_i calculates γ_i using Eq. (5.26)
11: t_i updates p_i using Eq. (5.27)

5.6 Numerical Results and A Case Study

In this section, we numerically analyze the proposed centralized schemes and conduct a case study to demonstrate the distributed scheme.

5.6.1 Simulation Settings

The studied monitoring network with transmission line model is shown in Figure 5.5. Without loss of generality, we assume that the length of the power line between two neighboring towers is d_0 km. However, the transmission towers are not necessarily installed in a straight line. To be more realistic, we assume that two neighboring towers have an angle of α to the east. For simplicity, the elevation of the terrain is not considered in the settings, and the transmission line is assumed to be installed on a flat plain. Additionally, we assume that each α_i is randomly chosen in $[-\pi/4, \pi/4]$. Practically, the sensors are deployed on the selected towers instead of all towers. In the simulation setting, we assume that the sensors are mounted at every k towers. An example with $k = 5$ is illustrated in Figure 5.5. Note that the wireless transmission distance between t_i and t_{i+1} is generally shorter than kd_0.

Without loss of generality, we assume that the multihop wireless sensor network (i.e. $\forall t_i \in \mathcal{T}_1$ and r_1) is deployed on the towers with an end-to-end power line length of 10 kilometers. The rest of the setting are as follows: for all t_i, we set $L_i = 200$ bit, $R_i = 200$ Kbps, $f_i = 1$ Hz. The end-to-end delay threshold $\bar{\tau} = 10$ ms, the processing gain

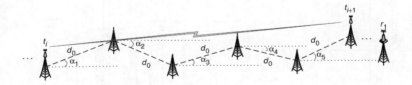

Figure 5.5 Simulation setting for transmission line.

$N = 64$, the packet length $M = 100$ bit, the physical layer transmission rate $R = 2$ Mbps, and the transmission frequency $f = 230$ MHz.

5.6.2 Comparison of the Centralized Schemes

Assuming that the sensors operate in the f=230 MHz spectrum, it is practical for them to transmit over several kilometers. Therefore, without loss of generality, we evaluate T from 1 to 10 if not otherwise specified. We first compare the results of $\mathcal{P}1$ and $\mathcal{P}1'$ with the following settings:

- The towers are deployed with randomly chosen α in the simulation setting.
- In each comparison with T sensors, we generate 100 random topologies.
- For each generated topology, the power allocations are calculated by both $\mathcal{P}1$ and $\mathcal{P}1'$.
- The computation time is recorded.

The comparison is shown in Figure 5.6. Each result in the figure is an average value of the 100 solutions generated from 100 random topologies (excluding the largest and the smallest results for further accuracy)

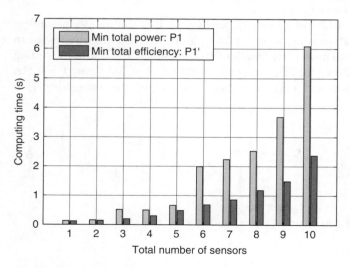

Figure 5.6 Computational time of $\mathcal{P}1$ and $\mathcal{P}1'$.

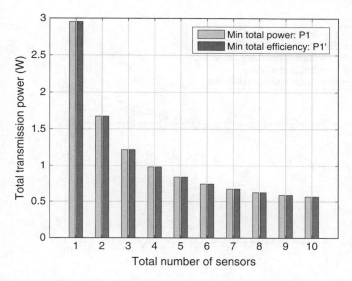

Figure 5.7 Total transmission power by solving $\mathcal{P}1$ and $\mathcal{P}1'$.

to the corresponding scheme. At we can see, solving $\mathcal{P}1'$ for the optimal solution is much faster than solving $\mathcal{P}1$. However, even when T is a small number (e.g. $T = 2, 3$), finding the optimal solution is too expensive to be applied for real-time operation of the monitoring system.

Second, we show the total transmission power for all the sensors with $T \in [1, 10]$. In order to show the impact on the number of sensors more precisely, we place the transmission towers in a straight line (i.e. $\alpha = 0$). As shown in Figure 5.7, the relaxed scheme $\mathcal{P}1'$ has almost the same power allocation results as scheme $\mathcal{P}1$.

Third, we set $T = 10$ and compare the results from different centralized schemes. Figure 5.8 shows the comparison of the normalized transmission efficiency of each sensor t_i with different centralized schemes. We first calculate the solution to $\mathcal{P}1$. The solution is then used as the benchmark value for the game theoretical scheme. In the game theoretical scheme, the weight factor is set as $k_i = 0.0000003$ for all t_i. As we can see, $\mathcal{P}1$ and $\mathcal{P}1'$ have results close to each other, which indicates that the relaxation of $\mathcal{P}1$ remains a good approximation. The game theoretical scheme also has a result close to that of $\mathcal{P}1$, because it is based on the benchmark setting from $\mathcal{P}1$. The equal link delay scheme requires linearly increasing transmission efficiency as the link gets closer to the gateway, because the amount of data increases

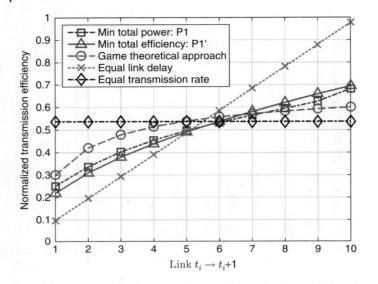

Figure 5.8 Normalized transmission efficiency of each sensor.

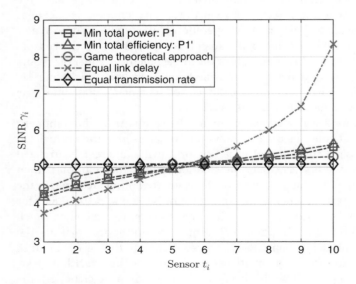

Figure 5.9 SINR of each sensor.

linearly as it is uploaded to the gateway. On the other hand, the equal transmission rate scheme results in a flat transmission efficiency.

Figure 5.9 shows the corresponding SINR γ_i for all t_i of the centralized schemes. Note that the last link requires a much higher SINR in the scheme with equal link delay; therefore the last sensor requires a much larger power supply component, which will make the deployment more complicated.

Figure 5.10 shows the delay τ_i for each link $t_i \rightarrow t_{i+1}$ achieved by the different centralized schemes. In comparison, all the schemes (except for the *equal link delay* one) have an increasing link delay as the sensors get closer to the gateway, because of the increase in aggregated data. The end-to-end delays for all the centralized schemes are the same as $\tau = \overline{\tau} = 10$ *ms*. The end-to-end delay for the game theoretical scheme is $\tau = 9.7660$ *ms* $< \overline{\tau}$. The game theoretical scheme achieves a slight advantage because of its slightly higher transmission power.

Figure 5.11 shows results of total transmission power achieved by different schemes. Compared to the optimal solution generated by $\mathcal{P}1$, $\mathcal{P}1'$ and game theoretical scheme have close-to-optimal results. For practical purpose, the equal transmission rate scheme is recommended over the equal delay scheme, since its results are closer to optimal.

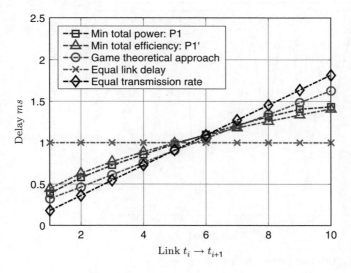

Figure 5.10 Delay of each link.

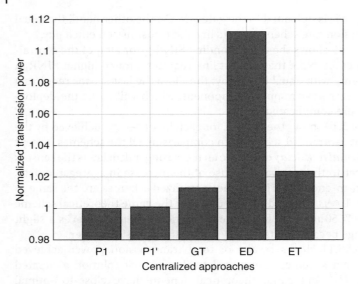

Figure 5.11 Comparison of normalized transmission power.

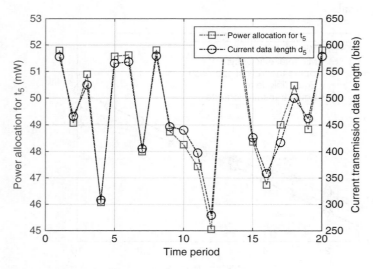

Figure 5.12 Dynamic power allocation for t_5.

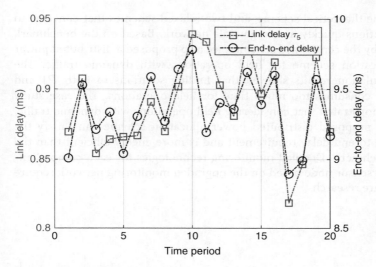

Figure 5.13 Corresponding τ_5 and τ.

5.6.3 Case Study

In the case study, the transmission line is randomly deployed following the previous setting with $T = 10$ sensors. It is then fixed throughout the discussion. At sensor t_i, its monitoring data l_i is assumed to be generated at a random rate $0 < \lambda_i < 1$, and the gathered data $0 \leq l_i \leq L_i$, $f_i = 1$ *Hz*. We first show the adaptive power allocation for sensor t_5 in 20 time slots $(20/f_i)$ to demonstrate its adaptability to the dynamic traffic in Figure 5.12. In Figure 5.13, we show the link delay τ_5 of link $t_5 \rightarrow t_6$ and the corresponding end-to-end delay τ. Note that the end-to-end delay is always below the required $\overline{\tau} = 10$ ms; therefore, the distributed power allocation scheme is valid for practical operations.

5.7 Summary

In this chapter, we designed and analyzed a wireless sensor–based monitoring network for the transmission lines in the smart gird. Specifically, we studied the power allocation for the wireless multihop sensor network between two OPGW gateways. In order to set the power allocation benchmark for the design of the wireless sensors, we proposed several centralized schemes, including two optimization problems, $\mathcal{P}1$ and $\mathcal{P}1'$, that minimize the total power consumption, a game theoretical scheme, and two relaxed schemes that converge to solutions quickly in a large scale network. Based on the benchmark set by the centralized schemes, we also proposed a distributed power allocation scheme for field operations with dynamic traffic. The simulation results showed that finding solutions to both $P1$ and $P1'$ consumes too much time for field operations. The case study demonstrated that considering field operations with dynamic traffic, our proposed distributed power allocation scheme can satisfy the end-to-end delay requirement and is more energy efficient than the benchmark. Detailed monitoring technologies for each comprehensive sensor node based on the upgraded monitoring network require future research.

6

A Real-Time Information-Based Demand-Side Management System

Demand response is a feature to be achieved in the smart grid. In this chapter, we study a few mechanisms to demonstrate the enhanced efficiency in the smart grid with demand response.

6.1 Background and Related Work

6.1.1 Background

Demand response (DR), also known as demand-side management system [27–29], utilizes real-time information in order to let the power grid generate and consume energy more efficiently while reducing fuel waste. A DR system is widely agreed to be effective at reducing the peak-to-average ratio (PAR) of energy consumption [28, 29, 98]. This improvement helps power suppliers reduce the extra fuel costs caused by dramatic and unpredictable margin fluctuations in power generation. Burning less fuel also helps reduce emission of greenhouse gas from those power generators. Moreover, since the control center gets the energy consumption schedule beforehand [83, 99–101], renewable sources such as photovoltaics (PV) and wind turbines, which are less stable and less controllable compared than conventional power generators, can support the power grid more efficiently. A higher proportion of such renewable sources will further reduce the burning of fuel by conventional power generators.

In this chapter, we first propose a centralized optimization problem $P1$ in order to reduce PAR to its minimum. Although a minimum PAR is obviously beneficial to the environment, it motivates neither power suppliers nor customers. Power suppliers especially must deploy and

Smart Grid Communication Infrastructures: Big Data, Cloud Computing, and Security,
First Edition. Feng Ye, Yi Qian, and Rose Qingyang Hu.
© 2018 John Wiley & Sons Ltd. Published 2018 by John Wiley & Sons Ltd.

maintain a more complicated cyber physical system than what AMI can offer to gather and distribute a huge amount of detailed information in real time. Therefore, a monetary incentive is needed to motivate power suppliers. Another centralized optimization problem $P2$ is then proposed to reduce the total cost of energy generation cost to power suppliers. Although power suppliers may be willing to adopt the DR system based on $P2$, it is based on *direct load control* (DLC) [99, 102, 103], which could be defective. In terms of communications, even if the massive centralized problem can be solved efficiently, the transmission overhead will create a huge burden on the communication network and require more advanced technical upgrades as well as more frequent maintenance. Moreover, customers could be reluctant to adopt such a DR system for two reasons. One reason is that the control center takes over the energy consumption scheduling from customers with no clear incentive for them to do so. The other reason is that DLC requires too much private information from customers.

To tackle those issues, we must have a DR system that clearly benefits customers, protects their privacy, and requires much less real-time information exchange compared with $P1$ or $P2$. We formulate a game with two approaches based on *smart pricing*, which is another major technique applied to the DR system. In one approach, customers get to compute the dynamic price based on their own load schedule, with the total load of the power grid given. In the other approach, the control center computes the price based on the total load of the power grid, and customers get that fixed price schedule that will not be affected by their local scheduling load. In either game theoretical approach, the payoff functions lead customer to more energy cost savings. Therefore, customers are motivated to adopt this DR system. In addition, customers reserve the right to control their power consumption, and by doing so, they keep their information private by submitting only the energy consumption schedule, which is always requested even in the traditional power grid. Moreover, since most of the calculation is performed locally at the customer side, the approaches are mostly distributed instead of centralized. The distributed game theoretical approach largely relieves the transmission overhead. More importantly, we prove that all $P1$, $P2$, and the game theoretical approach with locally computed dynamic smart pricing lead to the same minimum PAR. Therefore, while all parties get enough motivation to participate, the DR system can be deployed in a distributed way.

6.1.2 Related Work

DR in the smart grid has been studied by many researchers recently [27–29, 83, 98–101, 104–106]. However, most of the works focus on one of the parties (e.g. power suppliers when applying DLC or customers when adopting smart pricing) in the system only, without clarifying why others were overlooked. The game theoretical approach and smart pricing have also been widely adopted in most of the studies as efficient approaches. The works most relevant to this paper includes [28, 29, 99]. The authors of [28] proposed a noncooperative game played among residential customers, and a two-stage Stackelberg game theoretical approach where power suppliers as the leaders tend to maximize their profits and customers as the followers tend to minimize their costs. Since [28] mostly targeted residential customers, the benefits for other parties in the system were not clearly stated, and an impractical situation where the total load goes negative was carefully avoided. In [29], the authors proposed an efficient game theoretical approach for residential customers without a storage unit based on dynamic smart pricing to reduce the PAR. While computational efficiency was demonstrated, the global optimal PAR was not guaranteed by the distributed approach. The authors of [99] are among of the first to minimize PAR by using a distributed game theoretical approach among customers. Similarly, the consideration of a storage unit was reserved for their future work, and the global optimal PAR was not guaranteed.

We summarize the main contributions as follows:

- In order to benefit the entire society and the environment, we propose a DLC-based centralized approach to minimizing PAR.
- In order to show the benefits to the power suppliers, we propose another DLC-based centralized approach that minimizes the cost of power generation, and the power generation cost model considers all sources, including conventional, nonexpanding green energy, and expanding renewable energy sources.
- In order to motivate customers to adopt the DR system and protect their privacy, we propose smart pricing based game theoretical approaches that can maximize savings for customers adopting such a DR system. The game theoretical approaches are mostly distributed, and thus they alleviate the communication burden of the network.

- We prove that the proposed DLC-based approach and one of the smart-pricing-based distributed game theoretical approaches yield the same optimal solution (minimum PAR); therefore the distributed game theoretical approach can be applied in real situations while all parties observe clear benefits from the DR system.
- We provide extensive numerical analysis and simulation results to demonstrate our analysis. We also compare several distributed approaches in order to find the best way to deploy the DR system.

6.2   System Model

6.2.1   The Demand-Side Power Management System

The DR system under consideration mainly consists of three parts: the control center, power suppliers, and customers, as illustrated in Figure 6.1. Power generators include all major types, from fuel-consuming conventional power generators to renewable power generators. For simplicity, a micro grid that can be attached to or detached from the power grid is not considered, and customers do not have power generators. The control center is mainly responsible for power distribution. It also gathers data (e.g. energy consumption) and distributes control information (e.g. price, tariff, and emergency control signal). The information is available because of a two-way communication network in the DR system. At each customer site, there is a smart meter that is responsible for reporting the power

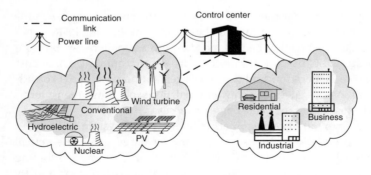

Figure 6.1 Demand-side power management system.

consumption and possible scheduling to the control center through the communication network. It is also responsible for receiving price, tariff, and other control information from the control center.

According to [107], major customers in the United States include residential, business, and industrial ones as shown in Figure 6.2. Those customers have different characteristics when consuming electric energy. For example, residential customers may consume power mostly from afternoon through midnight, business customers consume energy mostly during office hours, while industrial customers may have a longer peak consumption schedule because of continuous shifts by different work groups. The deployment of energy storage units and the use of plug-in hybrid vehicles (PHEV) are increasing rapidly. Such devices/appliances will increase energy consumption and change the current peak time schedule. For example, it is reasonable to assume that most customers will charge their PHEV during the night. For simplicity, the PHEV is considered as a hybrid of a normal energy-consuming appliance and an energy storage unit in the studied DR system.

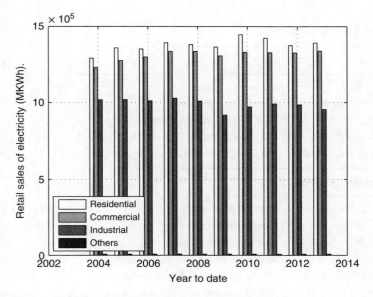

Figure 6.2 Power sale to the customers in United States.

6.2.2 Mathematical Modeling

Table 6.1 lists the key notations of sets and variables we use throughout the rest of this chapter. Let $\mathcal{N} = \{1, 2, \ldots, N\}$ be the set of all customers, where $N \triangleq |\mathcal{N}|$ is the total number of customers. Although the DR system can be modeled for any arbitrary time period to satisfy the assumptions, we consider a daily model in this work without loss of generality. Let one day be divided into several uniform time intervals, denoted as $\mathcal{T} = \{1, 2, \ldots, |\mathcal{T}|\}$.

Each customer n has a set of appliances $\mathbf{a}_n = \{a_n^1, \ldots, a_n^{A_n}\}$, where $A_n \triangleq |\mathbf{a}_n|$. Each appliance (e.g. a_n^i) has a daily energy consumption scheduling set $\mathbf{x}_n^i = \{x_n^i(t)|t \in \mathcal{T}\}$, which records the energy needed or consumed during each time interval. Moreover, each appliance a_n^i also has a predetermined energy requirement E_n^i for a time period (assuming one day for simplicity). Therefore, for a_n^i, it must satisfy

$$\mathbf{1}^T \mathbf{x}_n^i = E_n^i, \tag{6.1}$$

where $\mathbf{1}$ is a column vector of $1s$ and $(\cdot)^T$ calculates the transposition. Appliance a_n^i has an on/off operating scheduling set $\mathbf{t}_n^i = \{0, 1\}^{|\mathcal{T}|}$,

Table 6.1 Key sets and variables.

Global	
$\mathcal{N} = \{1, 2, \ldots, N\}$	set of customers
$\mathcal{T} = \{1, 2, \ldots, |\mathcal{T}|\}$	set of time intervals
$\mathbf{L} = \{L(t)|t \in \mathcal{T}\}$	daily energy consumption scheduling
$\mathbf{C} = \{C(L(t))|t \in \mathcal{T}\}$	set of total cost for each time interval
$\mathbf{p} = \{p(L(t))|t \in \mathcal{T}\}$	set of unit price for each time interval
Local (for customer n)	
$\mathbf{a}_n = \{a_n^1, a_n^2, \ldots, a_n^{A_n}\}$	set of appliances
$\mathbf{x}_n^i = \{x_n^i(t)|t \in \mathcal{T}\}$	daily energy scheduling set of a_n^i
$\mathbf{E}_n = \{E_n^i|i \le A_n\}$	Energy requirement set
$\mathbf{t}_n^i = \{0, 1\}^{|\mathcal{T}|}$	on/off operating scheduling set of a_n^i
$\mathbf{s}_n^- = \{s_n^-(t)|t \in \mathcal{T}\}$	discharging scheduling of storage unit
$\mathbf{s}_n^+ = \{s_n^+(t)|t \in \mathcal{T}\}$	charging scheduling of storage unit
$\mathbf{l}_n = \{l_n(t)|t \in \mathcal{T}\}$	daily energy consumption scheduling set

where 1 indicates that a_n^i is allowed to operate, whereas 0 indicates an off status for a_n^i. The on/off operating schedule can model the operating status more precisely than the model using an operating time period, which is more widely adopted [28, 29, 99]. For example, with an on/off operating schedule, it is possible to model a two-hour pause in an air-conditioning (AC) system. However, if it is modeled by the operating time period, the time period must be divided into three corelated sessions with extra constraints. With the on/off operating scheduling set $\mathbf{t}_n^i$, Eq. (6.1) can be rewritten into

$$(\mathbf{t}_n^i)^T \mathbf{x}_n^i = E_n^i. \tag{6.2}$$

For an appliance that needs to be used several times daily (e.g. a coffee machine), it can be modeled as multiple independent appliances with corresponding energy requirements and on/off schedules. For this reason, we want to emphasize that A_n may not necessarily be the exact number of appliances of customer n but the number that counts independent appliances.

When a_n^i is operating, its energy consumption is bounded by $\gamma_{n,i}^{\min}$ and $\gamma_{n,i}^{\max}$, and mathematically

$$\gamma_{n,i}^{\min} \mathbf{t}_n^i \leqslant \mathbf{x}_n^i \leqslant \gamma_{n,i}^{\max} \mathbf{t}_n^i. \tag{6.3}$$

Besides the appliances, let each customer (n) be equipped with an energy storage unit with a design capacity of $\bar{s}_n$. For simplicity, we assume that the storage unit has 100% discharging/charging efficiency, and the energy can be distributed for all the appliances within the power grid with 100% efficiency. In other words, the energy in storage units can support the appliances of customers themselves as well as be sold to the power grid. Let $\mathbf{s}_n^- = \{s_n^-(t)|t \in \mathcal{T}\}$ be the discharging scheduling set, and let $\mathbf{s}_n^+ = \{s_n^+(t)|t \in \mathcal{T}\}$ be the charging scheduling set. Like appliance energy consumption, the discharging/charging energy in each time interval is bounded by the safety thresholds $s_{n-}^{\max}/s_{n+}^{\max}$, which are expressed as

$$0 \leqslant \mathbf{s}_n^- \leqslant s_{n-}^{\max} \mathbf{1}, \tag{6.4}$$
$$0 \leqslant \mathbf{s}_n^+ \leqslant s_{n+}^{\max} \mathbf{1}. \tag{6.5}$$

To be more precise, the storage unit should not discharge and charge at the same time for efficiency, so we have

$$\mathbf{s}_n^- \circ \mathbf{s}_n^+ = \mathbf{0}, \tag{6.6}$$

where "∘" is the *entrywise/Hadamard product*, and $\mathbf{0}$ is the column vector with all 0s. Eq. (6.6) will help convert the storage unit model into a lossy one easily.

Let $\mathbf{s}_n = \{s_n(t)|t \in \mathcal{T}\}$ be the set of remaining energy at the beginning of each time interval,

$$
s_n(t) = \begin{cases} \tilde{s}_n, & t = 1 \\ s_n(t-1) - s_n^-(t-1) + s_n^+(t-1), & t = \in \mathcal{T}\backslash\{1\} \end{cases}
$$
(6.7)

where $\tilde{s}_n$ is the initial remaining capacity. Let $\bar{s}_n$ be the designed capacity of the storage unit; then

$$
0 \leq \tilde{s}_n \leq \bar{s}_n,
$$
(6.8)

$$
\mathbf{0} \preccurlyeq \mathbf{s}_n - \mathbf{s}_n^- + \mathbf{s}_n^+ \preccurlyeq \bar{s}_n \mathbf{1}.
$$
(6.9)

Let the daily energy consumption schedule for customer n be $\mathbf{l}_n = \{l_n(t)|t \in \mathcal{T}\}$. With the previous modeling, we now have

$$
\mathbf{l}_n = \sum_{i=1}^{A_n}(\mathbf{x}_n^i) - \mathbf{s}_n^- + \mathbf{s}_n^+.
$$
(6.10)

For the whole DR system, the global load schedule $\mathbf{L} = \{L(t)|t \in \mathcal{T}\}$ is calculated as

$$
\mathbf{L} = \sum_{i=1}^{N}\mathbf{l}_n.
$$
(6.11)

6.2.3 Energy Cost and Unit Price

Based on [108], power suppliers are categorized into three types: conventional generators using fossil fuels (e.g.coal), nonexpandable *green* sources (hydroelectric, nuclear), and expandable *renewable* sources (PV fields, wind farms). The net capacity of those generators/sources is shown in Figure 6.3. Conventional generators are still the major energy producers; however their proportion is decreasing. The proportion of nonexpandable generators is slowly decreasing since the total amount of energy is increasing. Although the proportion of expandable renewable energy is low, it is increasing at a faster pace. Therefore, we need to take into consideration all three categories of power suppliers for a more precise modeling.

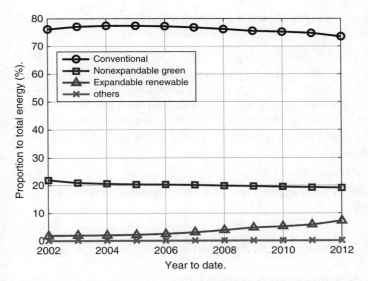

Figure 6.3 Existing net capacity by energy source and producer type [108].
Source: Data from http://www.eia.gov/electricity/annual/html/epa_03_01_a.html

- *Fuel-consuming conventional generators.* A quadratic cost function is widely adopted for these generators [27–29, 98, 99] as

$$C_c(l) = a_c l^2 + b_c l + c_c. \tag{6.12}$$

- *Nonexpandable green sources.* Assuming that the energy produced is predetermined by the fixed facilities, the total cost can be viewed as a fixed cost c_f plus a linear cost with respect to power transmission capacity as

$$C_f(l) = a_f l + c_f. \tag{6.13}$$

- *Expandable renewable sources.* Since most of their cost comes from the management of the facilities [109], by assuming the facilities can be on/off based on the load requirement, the total cost is increasing with respect to the load requirement. However it increases slower than that of the conventional generators [110], especially when *carbon tax* [111] applies. Therefore we adopt the following cost model for expandable renewable sources.

$$C_r(l) = a_r l \ln(l + 1) + c_r. \tag{6.14}$$

Taking into consideration the proportions of all the generators/sources, the overall energy cost is modeled as

$$C(l) = C_c(\beta_c l) + C_f(\beta_f l) + C_r(\beta_r l), \qquad (6.15)$$

where β_c, β_f and β_r are the proportions of the three power suppliers respectively, and $\beta_c + \beta_f + \beta_r = 1$. Let $\mathbf{C} = \{C(L(t))|t \in \mathcal{T}\}$ be the set of total costs for each time interval. At the customer side, the unit price (\$ per kWh) is more important than the total cost. For simplicity, let the energy be generated uniformly during a time period; the unit price is then calculated as

$$p(l) = \dot{C}(l) = C(l)/l. \qquad (6.16)$$

Finally let $\mathbf{p} = \{p(L(t))|t \in \mathcal{T}\}$ be the set of unit prices for each time interval.

6.3 Centralized DR Approaches

6.3.1 Minimize Peak-to-Average Ratio

One of the ultimate goals of applying DR is to reduce the peak-to-average ratio, which raises the first problem:

$$\mathcal{P}1 : \quad \min \frac{\sup_{t \in \mathcal{T}} \mathbf{L}}{\frac{1}{|\mathcal{T}|}(\mathbf{1}^T \mathbf{L})}, \qquad (6.17)$$

subject to constraints (6.4), (6.5), (6.6), (6.7), (6.8), (6.9), (6.10), $\forall n \in \mathcal{N}$ and constraint (6.11).

Lemma 6.1 Let $\mathbb{L}$ be the set of all possible daily load scheduling patterns ($\mathbb{L}$ is convex because of the convex, compact, and nonempty constraints). Then $\mathcal{P}1$ has a unique optimal solution $\mathbf{L}_1^* = \arg \min_{t \in \mathcal{T}}(\sup \mathbf{L}), \forall \mathbf{L} \in \mathbb{L}$.

Proof: First, since all appliances (including the storage units) consume energy from the power generators/sources and each appliance has a daily energy requirement, the daily total load $\mathbf{1}^T \mathbf{L} = \Gamma$ is observed as a constant. Therefore

$$\mathcal{P}1 \triangleq \mathcal{P}1.1 : \quad \min_{t \in \mathcal{T}} \sup \mathbf{L},$$

with the same constraints. Let the objective function for $\mathcal{P}1.1$ be $f(\mathbf{L}) = \sup_{t \in \mathcal{T}}\{L(t)|t \in \mathcal{T}\}$, which satisfies, for $0 \leq \theta \leq 1$,

$$f(\theta\mathbf{L}_1 + (1 - \theta)\mathbf{L}_2) = \sup_t(\theta L_1(t) + (1 - \theta)L_2(t))$$

$$\leq \theta \sup_t L_1(t) + (1 - \theta) \sup_t L_2(t)$$

$$= \theta f(\mathbf{L}_1) + (1 - \theta)f(\mathbf{L}_2). \tag{6.18}$$

Thus $f(\mathbf{L})$ is convex. Additionally, the constraint set is compact, convex, and nonempty. Therefore, $\mathcal{P}1.1$ has a unique solution $\mathbf{L}_1^\star$ and so does $\mathcal{P}1$. □

Note that although the existence of an optimal solution always stands, the uniqueness of the solution only stands when $\mathbf{L}$ is considered as the variable set, because multiple solutions to Eq. (6.11) may exist with a given $\mathbf{L}$. Although minimizing PAR is a desirable objective, it may not be convincing enough for power suppliers or customers to adopt such a DR system. For this reason, we further formulate another problem that has a monetary incentive as part of its objective function.

6.3.2 Minimize Total Cost of Power Generation

In the electricity energy market, power generators/sources are not yet fully competitive with each other, since some of the technologies are still too expensive to apply and they operate based on a government subsidy [112]. Moreover, sophisticated regulatory mechanisms are needed to avoid arbitrarily high prices produced by monopolies and rigid electric energy demand. Therefore, we focus on a cost-oriented instead of a profit-oriented objective. In short, $\mathcal{P}2$ minimizes the total cost to the power suppliers, such that

$$\mathcal{P}2: \quad \min \, \mathbf{p}^T\mathbf{L}, \tag{6.19}$$

subject to constraints (6.4), (6.5), (6.6), (6.7), (6.8),

(6.9), (6.10), $\forall n \in \mathcal{N}$ and constraint (6.11).

Lemma 6.2 $\mathcal{P}2$ has a unique optimal solution $\mathbf{L}_2^\star = \arg\min \mathbf{p}^T\mathbf{L}$, $\forall \mathbf{L} \in \mathbb{L}$.

Proof: The objective function of $\mathcal{P}2$ is

$$\mathbf{p}^T\mathbf{L} = \sum_t (p(L(t))L(t)) = \sum_{t\in\mathcal{T}} C(L(t)). \tag{6.20}$$

For simplicity, let $x \triangleq L(t)$ be the argument in this proof. It is obvious that $C(x)$ is increasing and strictly convex. Since Eq. (6.20) is a composition of $C(x)$, thus $\sum_{t\in\mathcal{T}}(p(L(t)) \cdot L(t))$ is strictly convex with respect to $L(t)$. With the same compact, convex, and nonempty constraint set compared to $\mathcal{P}1$, $\mathcal{P}2$ thus has a unique optimal solution. □

Let function $g(\mathbf{L}) \triangleq \mathbf{p}^T\mathbf{L}$. Then the optimal solution to $\mathcal{P}2$ can be found by solving the necessary and sufficient conditions of KKT [113].

Lemma 6.3 Let v be the Lagrange multiplier of the equality constraint of $\mathcal{P}2$ ($\mathbf{1}^T\mathbf{L} = \Gamma$), and let $v^\star$ minimize the dual problem of $\mathcal{P}2$ over v. The optimal solution $\mathbf{L}_2^\star = \{L_2^\star(t)|t \in \mathcal{T}\}$ to $\mathcal{P}2$ is calculated as

$$\begin{cases} L_2^\star(t) = \max\left\{ -\sum_{n\in\mathcal{N}} s_{n-}^{\max}, \underset{L(t)}{\arg}\left(\frac{\partial g(\mathbf{L})}{\partial L(t)} + v^\star \right) \right\}, \ \forall t \in \mathcal{T} \\ v^\star = \underset{v}{\arg}\left(\sum_{t\in\mathcal{T}} L_2^\star(t) = \Gamma \right). \end{cases}$$
$$\tag{6.21}$$

Note that solution presented in Eq. (6.21) is unique with respect to $\mathbf{L}$, and it may have multiple solutions with respect to the detailed energy consumption scheduling patterns to all the appliances.

Lemma 6.4 If $\mathbf{L}^\star$ is the optimal solution to $\mathcal{P}1$, it is also the optimal solution to $\mathcal{P}2$.

Proof: Let $\mathbf{x} = \{x_t|t \in \mathcal{T}\} \triangleq \mathbf{L}_1^\star$ be the optimal solution of $\mathcal{P}1$, and let $\mathbf{y} = \{y_t|t \in \mathcal{T}\} \triangleq \mathbf{L}_2^\star$ be the optimal solution of $\mathcal{P}2$ in this proof. Also, reorganize $\mathbf{x}, \mathbf{y}$ to be nondescending sets such that $x_i \leq x_{i+1}$, $y_i \leq y_{i+1}$, $i \in \mathcal{T}$ to provide a clearer illustration. Note that $\sup_t \mathbf{x} = x_{|\mathcal{T}|}$, and $\sup_t \mathbf{y} = y_{|\mathcal{T}|}$. We then prove this lemma by contradiction. Assuming $\mathbf{x} \neq \mathbf{y}$, then it must be

$$x_{|\mathcal{T}|} < y_{|\mathcal{T}|}, \tag{6.22}$$

$$\sum_{i\in\mathcal{T}} x_i p(x_i) > \sum_{i\in\mathcal{T}} y_i p(y_i). \tag{6.23}$$

Furthermore, inequality (6.23) indicates that

$$\sum_{i \in \mathcal{T} \backslash \{|\mathcal{T}|\}} (x_i p(x_i) - y_i p(y_i)) > y_{|\mathcal{T}|} p(y_{|\mathcal{T}|}) - x_{|\mathcal{T}|} p(x_{|\mathcal{T}|}). \tag{6.24}$$

The left hand side of inequality (6.24) is maximized when $x_i = y_i = 0$, $i \in \mathcal{T} \backslash \{|\mathcal{T}|\}$ since $\mathbf{1}^T \mathbf{x} = \mathbf{1}^T \mathbf{y} = \mathbf{1}^T \mathbf{L} = \Gamma$ is the daily total load and function $g(x) = xp(x)$ is increasing and strict convex with respect to x, we have

$$(\Gamma - x_{|\mathcal{T}|}) p(\Gamma - x_{|\mathcal{T}|}) - (\Gamma - y_{|\mathcal{T}|}) p((\Gamma - y_{|\mathcal{T}|}))$$

$$\geq \sup_{\mathbf{x}, \mathbf{y} \in \mathbb{L}} \left(\sum_{i \in \mathcal{T} \backslash \{|\mathcal{T}|\}} (x_i p(x_i) - y_i p(y_i)) \right). \tag{6.25}$$

However, because of inequality (6.22), then

$$x_{|\mathcal{T}|-1} \leq \Gamma - x_{|\mathcal{T}|} < x_{|\mathcal{T}|},$$
$$y_{|\mathcal{T}|-1} \leq \Gamma - y_{|\mathcal{T}|} < y_{|\mathcal{T}|}, \tag{6.26}$$

and the fact that

$$\Gamma - x_{|\mathcal{T}|} - \Gamma - y_{|\mathcal{T}|} = y_{|\mathcal{T}|} - x_{|\mathcal{T}|}. \tag{6.27}$$

We must then have

$$\sup_{\mathbf{x}, \mathbf{y} \in \mathcal{L}} \sum_{i \in \mathcal{T} \backslash \{|\mathcal{T}|\}} (x_i p(x_i) - y_i p(y_i)) \leq (\Gamma - x_{|\mathcal{T}|}) p(\Gamma - x_{|\mathcal{T}|})$$

$$- (\Gamma - y_{|\mathcal{T}|}) p(\Gamma - y_{|\mathcal{T}|})$$

$$\leq y_{|\mathcal{T}|} p(y_{|\mathcal{T}|}) - x_{|\mathcal{T}|} p(x_{|\mathcal{T}|}). \tag{6.28}$$

Note that inequality (6.28) contradicts inequality (6.24), and thus $\mathbf{x} = \mathbf{y}$ (or $\mathbf{L}_1^\star = \mathbf{L}_2^\star = \mathbf{L}^\star$ after proper ordering), which completes the proof. □

Theorem 6.1 Minimizing the total cost reaches the minimum PAR: $\mathcal{P}2 \triangleq \mathcal{P}1$.

Although power suppliers may be willing to adopt a DR system according to $\mathcal{P}2$ so that they can reduce their total cost, it still has three major issues to solve for either $\mathcal{P}1$ or $\mathcal{P}2$. Customer privacy is the first issue. Since both $\mathcal{P}1$ and $\mathcal{P}2$ are executed exclusively at the control center side (often regarded as DLC schemes), all customers must submit detailed information to the control center and willingly let the control center schedule their power usage. The incentive for the customers to adopt a DR system is the second issue. Although a DR system smooths PAR and reduces the total cost, it is not clear whether customers can benefit from $\mathcal{P}1$ or $\mathcal{P}2$ or not. Third, both $\mathcal{P}1$ and $\mathcal{P}2$ are centralized optimization problems, which can get quite complicated and computation-intensive to solve. Even if the problems can be solved efficiently, the huge overhead of raw data gathering puts too much burden on the communication network. Therefore, we need a distributed DR system that also protects the customers by letting them control their own appliances.

6.4 Game Theoretical Approaches

6.4.1 Formulated Game

Smart pricing is another widely adopted cost strategy in order to attract customers' interest. Moreover, a game theoretical approach is an efficient way to solve a problem in a distributed fashion with some limited sharing of information. Therefore, we formulate a non-cooperative game $\mathcal{G} = [\mathcal{N}, \{\mathbb{L}_i\}, \{h_i(\cdot)\}]$, where $\mathbb{L}_i$ is the strategy set (all possible load scheduling patterns) of player (customer hereafter for consistency) i (the notation of a customer is changed from n to i, which is more generic for game theoretical approach), and $h_i(\cdot)$ is the payoff function of customer i, which is

$$h_i(\mathbf{l}_i) = (\mathbf{\Omega}^T - \mathbf{p}^T)\mathbf{l}_i, \tag{6.29}$$

where $\mathbf{\Omega}$ is a flat-rate price vector if a smart pricing strategy is not applied. This payoff function shows the saving of a customer from adopting this particular DR system. Intuitively, if adopting a DR system reduces their cost, customers are willing to participate. In this game, each customer calculates a best response $\mathbf{l}_i^\star$ with a given $\mathbf{l}_{-i} = \{\mathbf{l}_1, \ldots, \mathbf{l}_{i-1}, \mathbf{l}_{i+1}, \ldots, \mathbf{l}_N\}$ such that

$$h_i(\mathbf{l}_i^\star, \mathbf{l}_{-i}) \geq h_i(\mathbf{l}_i, \mathbf{l}_{-i}), \ \forall \mathbf{l}_i \in \mathcal{L}_i. \tag{6.30}$$

The best response in Eq. (6.30) is the same as calculating

$$\mathbf{l}_i^\star = \arg\max_{\mathbf{l}_i \in \mathbb{L}_i} h_i(\mathbf{l}_i, \mathbf{l}_{-i}), \tag{6.31}$$

subject to constraints (6.4), (6.5), (6.6), (6.7), (6.8), (6.9), (6.10).

Note that Eq. (6.31) has a specific meaning for its incentive, which is to find the set of load scheduling patterns that maximize the energy cost saving for customer i by adopting a DR system.

Since each customer has relatively stable daily energy consumption levels, that is, $\sum_t l_i(t) = \Gamma_i$, $\forall i \in \mathcal{N}$, the flat-rate price yields a constant cost for each customer. Therefore, the problem in Eq. (6.31) equals the one as follows:

$$\mathbf{l}_i^\star = \arg\min_{\mathbf{l}_i \in \mathbb{L}_i} \mathbf{p}^T \mathbf{l}_i, \tag{6.32}$$

subject to constraints (6.4), (6.5), (6.6), (6.7), (6.8), (6.9), (6.10).

6.4.2 Game Theoretical Approach 1: Locally Computed Smart Pricing

In this approach, we assume that each customer submits an initial load schedule $\mathbf{l}_i^0 = \{l_i^0(t) | t \in \mathcal{T}\}$, $\forall i \in \mathcal{N}$ to the control center and then the control center broadcasts the initial total load schedule $\mathbf{L}^0 = \{L^0(t) | t \in \mathcal{T}\} = \sum_{i \in \mathcal{L}} \mathbf{l}_i^0$ to the customers. In this approach, the price computing function in Eq. (6.16) is known to all customers. Then each customer will be able to compute the smart pricing schedule as

$$p_i(l_i(t)_i | L^0(t), l_i^0(t)) = p_i(L^0(t) - l_i^0(t) + l_i(t)), \forall t \in \mathcal{T}. \tag{6.33}$$

Let $\mathbf{p}_i = \{p_i(L^0(t) - l_i^0(t) + l_i(t)) | t \in \mathcal{T}\}$. Customer i will need to solve the following problem to find $\mathbf{l}_i^\star$,

$$\mathcal{P}3: \quad \min_{\mathbf{l}_i \in \mathcal{L}_i} \mathbf{p}_i^T \mathbf{l}_i, \tag{6.34}$$

subject to constraints (6.4), (6.5), (6.6), (6.7), (6.8),

(6.9), (6.10).

Lemma 6.5 With given $\mathbf{l}_i^0$ and $\mathbf{L}^0$, $\mathcal{P}3$ has a unique optimal solution with respect to $\mathbf{l}_i = \{l_i(t) | t \in \mathcal{T}\}$.

Proof: The objective function of $\mathcal{P}3$ is analogous to that of $\mathcal{P}2$ and thus is monotonically increasing and strictly convex. The constraint set is also compact, convex, and nonempty. Thus $\mathcal{P}3$ has a unique optimal solution. □

Let function $g(\mathbf{l}_i) = \mathbf{p}_i^T \mathbf{l}_i$, and Γ_i be the total energy requirement for customer i. Then the optimal solution to $\mathcal{P}3$ can be found by solving the necessary and sufficient conditions of KKT [113].

Lemma 6.6 Let v_i be the Lagrange multiplier of the equality constraint, and $v_i^\star$ minimizes the dual problem of $\mathcal{P}3$ over v_i. The optimal solution $\mathbf{l}^\star = \{l^\star(t)|t \in \mathcal{T}\}$ to $\mathcal{P}3$ is

$$
\begin{cases}
l^\star(t) = \max\left\{ -s_{i-}^{\max}, \underset{l(t).}{\arg}\left(\dfrac{\partial g(\mathbf{l})}{\partial l(t)} + v^\star \right) \right\}, \quad \forall t \in \mathcal{T} \\
v_i^\star = \underset{v_i}{\arg}\left(\sum_{t \in \mathcal{T}} l^\star(t) = \Gamma_i \right).
\end{cases}
\tag{6.35}
$$

Definition 6.1 *Nash equilibrium (NE)* a scheduling set $\mathbf{l}^\star = (\mathbf{l}_1^\star, \mathbf{l}_2^\star, \dots, \mathbf{l}_N^\star)$ is an NE of $\mathcal{G} = [\mathcal{N}, \{\mathbb{L}_i\}, \{h_i(\cdot)\}]$ if, $\forall i \in \mathcal{N}$, $\forall \mathbf{l}_i \in \mathbb{L}_i$, $h_i(\mathbf{l}_i^\star, \mathbf{l}_{-i}^\star) \geq h_i(\mathbf{l}_i, \mathbf{l}_{-i}^\star)$, where $\mathbf{l}_{-i}^\star = (\mathbf{l}_1^\star, \dots, \mathbf{l}_{i-1}^\star, \mathbf{l}_{i+1}^\star, \dots, \mathbf{l}_N^\star)$.

Lemma 6.7 $\mathcal{G}$ has a unique NE by $GA1$.

Proof: First, the payoff function (Eq. (6.29)) is strictly concave, and the constraint set for this approach is compact, convex, and nonempty; thus NE exists [28, 114]. Second, lemma 6.5 guarantees that the best response of each player can be found uniquely. Therefore NE exists uniquely to $GA1$. □

Lemma 6.8 Algorithm 6.1 converges to NE.

Proof: According to Eq. (6.30), the payoff of each customer increases after each iteration. Because the payoff is bounded, the algorithm will converge to an equilibrium, which is the NE. □

Algorithm 6.1 is proposed to approach the NE of $GA1$.

Algorithm 6.1 Algorithm to find NE by $GA1$

Input: γ_i, $\mathbf{t}_i$, $\mathbf{E}_i$, $s_{i-}^{\max}$, $s_{i+}^{\max}$, $\bar{s}_i$, $\forall i \in \mathcal{N}$, $\mathcal{T}$
Output: $\mathbf{l}_i^\star$, $\forall i \in \mathcal{N}$;
 1: Each player computes a feasible $\mathbf{l}_i^0$
 2: **while** NE is not achieved **do**
 3: $\mathbf{L}^0 \leftarrow \sum_{i \in \mathcal{N}} \mathbf{l}_i^0$; // Computed and distributed by the control center
 4: $\mathbf{l}_i^0 \leftarrow \mathbf{l}_i^\star = \arg \mathcal{P}3$, $\forall i \in \mathcal{N}$; // Computed by customer i
 5: **end while**
 6: $\mathbf{l}_i^\star \leftarrow \mathbf{l}_i^0$, $\forall i \in \mathcal{N}$; // Output NE

Theorem 6.2 The NE of $\mathcal{G}$ by $GA1$ is also the optimal solution to $\mathcal{P}2$.

Proof: According to the definition of the best response, we have

$$\sum_{i \in \mathcal{N}} h_i(\mathbf{l}_i^{\star}, \mathbf{l}_{-i}) \geq \sum_{i \in \mathcal{N}} h_i(\mathbf{l}_i, \mathbf{l}_i), \ \forall \mathbf{l}_i \in \mathbb{L}_{-i},$$

$$\Rightarrow \sum_{i \in \mathcal{N}} g(\mathbf{l}_i^{\star}, \mathbf{l}_{-i}) \leq \sum_{i \in \mathcal{N}} g(\mathbf{l}_i, \mathbf{l}_{-i}), \ \forall \mathbf{l}_i \in \mathbb{L}_i,$$

$$\Rightarrow \sum_{i \in \mathcal{N}} g(\mathbf{l}_i^{\star}, \mathbf{l}_{-i}^{\star}) \leq \mathbf{P}^T \mathbf{L}, \quad \forall \mathbf{L} \in \mathbb{L},$$

$$\Rightarrow \mathbf{p}_i(\mathbf{l}_i^{\star}, \mathbf{l}_{-i}^{\star}) \sum_{i \in \mathcal{N}} \mathbf{l}_i^{\star} \leq \mathbf{P}^T \mathbf{L}, \quad \forall \mathbf{L} \in \mathbb{L},$$

$$\Rightarrow \sum_{i \in \mathcal{N}} (\mathbf{l}_i^{\star}) = \arg \min_{\mathbf{L} \in \mathbb{L}} \mathbf{P}^T \mathbf{L}. \tag{6.36}$$

$\square$

Theorem 6.2 demonstrates that $GA1$ not only favors the customers but also minimizes the total cost and thus favors power suppliers as well. So $GA1$ minimizes PAR according to theorem 6.1. However, the control center may not want to release the price-calculating function to customers. We then propose another approach based on a precalculated fixed-price schedule.

6.4.3 Game Theoretical Approach 2: Semifixed Smart Pricing

In this approach, each customer i also submits an initial load schedule $\mathbf{l}_i^0$; however, the price function is hidden from customers. Only the control center is able to calculate the fixed price vector with each $\mathbf{L}^0$ as

$$\tilde{\mathbf{p}} = \{p(L^0(t)) | t \in \mathcal{T}\}. \tag{6.37}$$

Then each customer i will solve the following problem to find $\mathbf{l}_i^{\star}$,

$$\mathcal{P}4: \quad \min_{\mathbf{l}_i \in \mathcal{L}_i} \tilde{\mathbf{p}}^T \mathbf{l}_i, \tag{6.38}$$

so that constraints (6.4), (6.5), (6.6), (6.7), (6.8), (6.9), (6.10).

Note that $\mathcal{P}4$ is a linear optimization problem, which can be solved with a unique solution. However, the NE for $\mathcal{G}$ is not guaranteed, since the objective function is no longer strictly concave.

6.4.4   Mixed Approach: Mixed $GA1$ and $GA2$

The mixed approach is to apply both $GA1$ and $GA2$ approaches based on the properties of customers. As mentioned earlier, $GA1$ reveals the price function to the customers, which may not favor the control center. The total load is also given to all customers. However, customers who use a significant proportion of energy may not want to reveal such private information. $GA2$ does not have those issues but it does not guarantee the optimal solution of $P1$ or $P2$. In the mixed approach, large-energy-consumption customers adopt $GA1$ (e.g. business and industrial customers), and regular-energy-consumption customers (e.g. residential customers) adopt $GA2$. In the numerical analysis and simulation section, we will show that the mixed approach converges to the minimum PAR in practice.

6.5   Precision and Truthfulness of the Proposed DR System

Because customers could have exceptional energy consumption, a one-time scheduling approach [28, 29, 99] can hardly be followed strictly. In order to increase the precision of the proposed DR system, the system should run at the beginning of each time interval for the rest of the day. The control center and the computing device of each customer (e.g. the smart meter) should save the previous status and subtract it from the constraints when calculating the load scheduling for the rest of the day. With the DR system following schedule, and since the payment is collected after each time interval based on the real energy usage in that interval, for $P1$, $P2$ and $GA1$, the minimum PAR, the minimum total cost, and the maximum local saving will be achieved only when customers report the load schedules truthfully. The conclusion is quite intuitive because lying about the load will keep customers from reaching the optimal solution. Therefore, the truthfulness of the proposed DR systems should be guaranteed.

6.6   Numerical and Simulation Results

6.6.1   Settings

Assume that power suppliers are available to support customers with any energy requirement. The control center is able to gather information from and distribute it to the power suppliers and the customers in real time (e.g. 100 *ms*). The customers are categorized into residential, business, and industrial types. Without loss of generality, we assume that the daily energy consumption of the customers follows the proportions obtained from the data in Figure 6.2. Specifically, as shown in Figure 6.4, residential customers consist of three types (i.e. families in big houses, families in townhouses, and families in apartments), each with 55 kilowatt-hours, 41 kilowatt-hours, and 33 kilowatt-hours daily average energy consumption respectively. Each residential customer type has 50 customers. Business customers have two types (i.e. day-time based business and shopping malls), each with 2400 kilowatt-hours and 2700 kilowatt-hours daily average energy consumption respectively, and each type has one customer. Industrial customers have two types (i.e. nonstop shift-based manufacturers and day-time based industry), each with 2100 kilowatt-hours and 2500 kilowatt-hours daily average energy consumption respectively, and each type has one customer. Note that the settings for the customers are flexible as long as the total energy consumption of each category follows the practical data.

The granularity of time intervals is important to the DR system. As shown in Figure 6.5 and Figure 6.6, when $|\mathcal{T}| = 24$, the total load

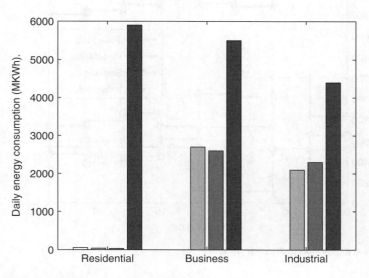

Figure 6.4 Daily consumption of customers.

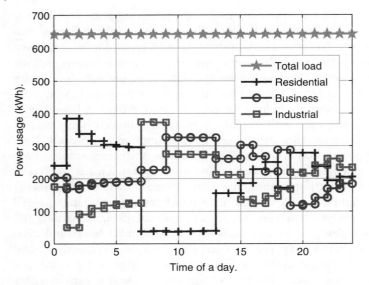

Figure 6.5 Solution to $\mathcal{P}1$ with $|\mathcal{T}| = 24$.

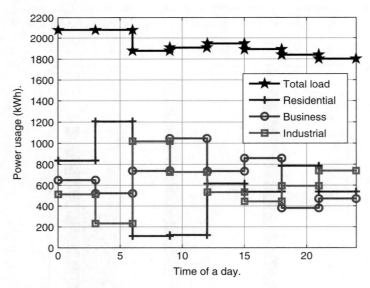

Figure 6.6 Solution to $\mathcal{P}1$ with $|\mathcal{T}| = 8$.

Table 6.2 Residential settings for the case study.

App a	E_n^a (kWh)	$\gamma_{n,j}^{min}$	$\gamma_{n,j}^{max}$	Start	End
Residential					
Storage	$1/6 \sum E_n^a$	$-E_n^a/3$	$E_n^a/3$	0	24
PHEV	18, 18, 9	0	$E_n^a/3$	0	8
AC1	5, 3, 2	$E_n^a/48$	$E_n^a/16$	0	24
AC2	5, 3, 2	$E_n^a/4$	$E_n^a/4$	14	18
Cooking	2, 1, 1	0	$E_n^a/1$	17,17,18	20,20,20
Dish washer	2, 1, 1	0	$E_n^a/2$	20,20,20	24,23,22
Washing/dryer	5, 3, 2	0	$E_n^a/2$	20,20,18	24,23,22
Electronics	2, 2, 2	$E_n^a/8$	$E_n^a/3$	18	24
Others	4, 2, 1	$E_n^a/48$	$E_n^a/12$	0	24

of the power grid is constant throughout the day, while it fluctuates when $|\mathcal{T}| = 8$.

Some of the detailed initial settings for residential customers are shown in Table 6.2. We assume that each storage unit is able to hold 1/6 of the daily power usage, each family has 2 PHEVs for type 1 and type 2 residential customers, and customers living in apartments have 1 PHEV each. Each PHEV holds 9 kilowatt-hours of electricity. The power usage and rough schedules for other appliances are estimates based on day-to-day experiences. For simplicity, the on/off schedule is shown as start/end times. Note that some of the appliances are shown as multiple ones to make the modeling more precise; for example, AC for residential users is shown as ventilation all day and fully operating in the afternoon.

6.6.2 Comparison of $\mathcal{P}1$, $\mathcal{P}2$ and $GA1$

In this subsection, we evaluate the three DR mechanisms, including the two centralized ones and the game theoretical approach where smart pricing is computed locally. Figure 6.7 shows that the $\mathcal{P}1$, $\mathcal{P}2$, and $GA1$ all reach the same optimal result with respect to the total load. However, each customer receives a different load scheduling. This is observed in all three DR systems.

To better illustrate the results from the three DR mechanisms, more detailed results for residential customers are shown in Figure 6.8.

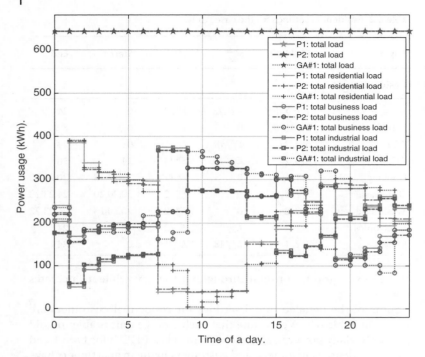

Figure 6.7 Load schedules by $P1$, $P2$ and $GA1$.

Specifically, Figure 6.8(a), Figure 6.8(c), and Figure 6.8(c) show the results for three types of residential customers respectively. For each type of residential customer, the results are given for eight hours only. Note that $GA1$ leads to a negative load for some customers, as shown in Figure 6.8(c). In the smart grid, some customers sell extra energy from their storage unit when the price is high to maximize their savings.

Similarly, load schedules for business customers are shown in Figure 6.9. Load schedules for industrial customers are shown in Figure 6.10.

6.6.3 Comparison of Different Distributed Approaches

Figure 6.11 shows the load scheduling results of $GA1$, $GA2$, mixed GA, and a distributed approach when all customers intend to minimize their local PAR. From the simulation results, we can see that both $GA1$

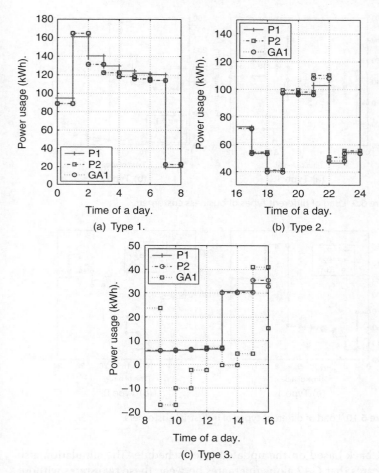

Figure 6.8 Load of different types of residential customers.

and mixed *GA* converge to the optimal PAR, while neither *GA2* nor *min local PAR* minimizes PAR.

In fact, *GA2* performs the worst in the simulation when all customers are considered. Because of the fixed smart price, customers will shift their load to the time intervals with low prices without considering the consequences of doing so. When all customers do so, the time intervals with a relative low previous load will be scheduled with a higher load, and the price will go up. Then customers will

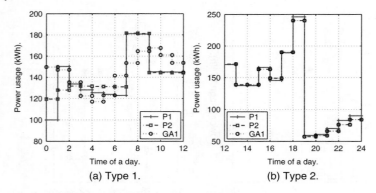

Figure 6.9 Load of different types of business customers.

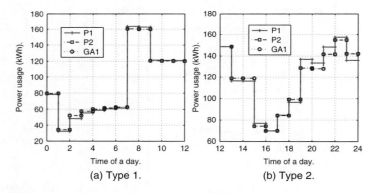

Figure 6.10 Load of different types of industrial customers.

shift back based on the updated price schedule. The simulation also indicates that $GA2$ alone fluctuates between these two states without converging to the NE. Figure 6.12 shows the two states of the total load schedule by $GA2$ for the first eight time intervals. Figure 6.13, Figure 6.14, and Figure 6.15 show the average load schedules by $GA1$ and mixed GA for the first type of customers in each category. It appears that $GA1$ schedules a smoother load for each customer, because residential customers have better assessments of the cost changes based on their updated schedule with $GA1$. When applying mixed GA, although each customer does not affect the price much, they together still cause a major impact. Therefore, both business and

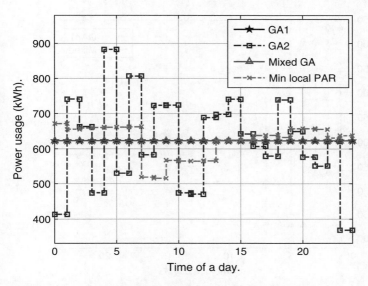

Figure 6.11 Load scheduling from different distributed approaches.

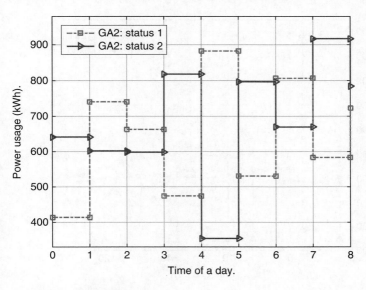

Figure 6.12 *GA2* illustration.

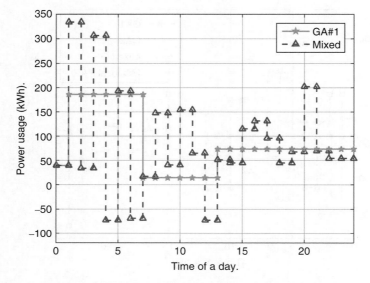

Figure 6.13 Load achieved by *GA* for type 1 residential customers.

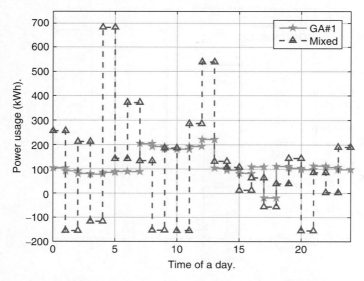

Figure 6.14 Load achieved by *GA* for type 1 business customers.

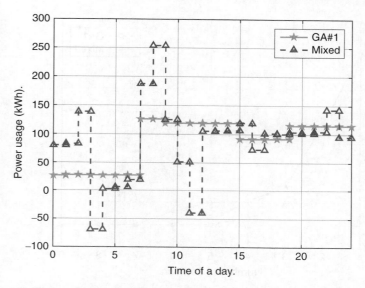

Figure 6.15 Load achieved by *GA* for type 1 industrial customers.

industrial customers must schedule their loads with more fluctuation to adapt to the residential load.

Figure 6.16 shows that both *GA*1 and mixed *GA* converge quickly to the NE. In game theoretical approaches, the uplink to the network transmits the load schedule of each customer to the control center, and the control center broadcasts the total load schedule to all customers. Although it requires multiple iterations, it also requires much less data transmission compared with the centralized approach, which needs much more detailed data from the customers, and the scheduled appliance load must be delivered to each customer individually. Moreover, it is worth mentioning that energy providers may not want to declare the true cost function to all the customers in practice. With the adoption of mixed GA, only a few customers are required to know the cost function, and some confidential agreements can be made.

6.6.4 The Impact from Energy Storage Unit

Whether to include energy storage units in the smart grid may contribute to DR in the smart grid. From Figure 6.17 we can see that the total load schedule is no longer a straight line without the assistance

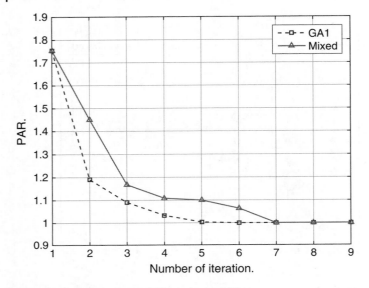

Figure 6.16 Convergence of *GA*1 and mixed *GA*.

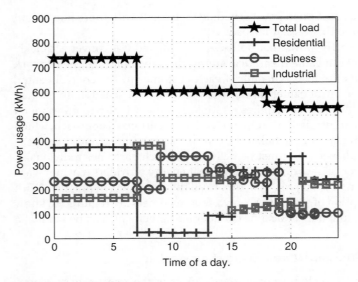

Figure 6.17 Load schedule without storage units.

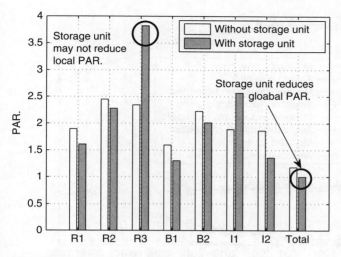

Figure 6.18 Impact of storage units on PAR.

of storage units. Therefore, storage units on the customers' side will certainly improve the efficiency of DR mechanisms. In theory, if the capacity of the energy storage units is unlimited, the smart grid would have enough buffer to store all electricity that is overgenerated, especially that from renewable power sources. In practice, customer may use electric vehicles as energy storage units in the future.

Without energy storage units, the PAR for the power grid would be higher, as shown in Figure 6.18. However, storage units may not necessarily help reduce the PAR for each customer individually.

6.6.5 The Impact from Increasing Renewable Energy

In the future, the smart grid will integrate more renewable energy sources. Based on the data shown Figure 6.3, we can estimate the different power suppliers up to the year 2020. As shown in Figure 6.19, renewable energy sources will be more than 20% of the smart grid.

Based on these estimates, the total energy cost to te customers in 2014, 2018, and 2020 is shown in Figure 6.20. Clearly, the expansion of renewable energy would help to reduce energy costs.

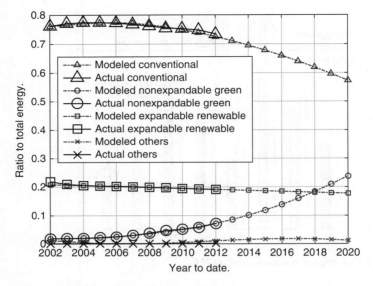

Figure 6.19 An estimate of different power suppliers.

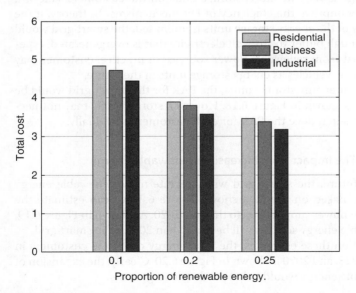

Figure 6.20 Corresponding total cost estimates for future years.

6.7 Summary

In this chapter, we studied demand response in the smart grid. In order to motivate and benefit all parties, including society (the environment), power suppliers, and customers, we proposed several approaches for the DR system. First, $\mathcal{P}1$ directly minimizes PAR. Second, a DLC-based cost minimization approach $\mathcal{P}2$ motivates the power suppliers. However, both $\mathcal{P}1$ and $\mathcal{P}2$ fail to protect the privacy of customers, and their communication overhead is too high for deployment in a real-time system. We proposed two further smart pricing-based game theoretical approaches $GA1$ and $GA2$ to address the shortcomings of $\mathcal{P}1$ and $\mathcal{P}2$. We successfully proved that both $\mathcal{P}2$ and $GA1$, in which each customer calculates the dynamic price locally, reach the solution to $\mathcal{P}1$ (min PAR). In the numerical analysis and the simulations, we further demonstrated the results and compared the performance of several distributed approaches.

7

Intelligent Charging for Electric Vehicles—Scheduling in Battery Exchanges Stations

An environmentally friendly lifestyle is the trend in recent years, as the number of plug-in hybrid electric vehicles (PHEVs) and electric vehicles (EVs) increases each day. As discussed in the previous chapter, PHEVs and EVs may also be considered as energy storage units in the smart grid to contribute in demand response. In this chapter, we study intelligent smart charging schemes for PHEVs and EVs in the smart grid. In particular, we introduce battery exchange stations into the demand-side management system so that the PHEVs and EVs can help smooth the load on the power grid. Two centralized schemes for optimal charging are proposed to achieve the minimum peak-to-average ratio in the smart grid. We also propose a game theoretical scheme so that the battery exchange stations can participate in the DSM system without handing over control to the control center.

7.1 Background and Related Work

7.1.1 Background and Overview

PHEVs and EVs (we will use PHEV in the rest of the chapter for simplicity) consume much less fossil fuel and produce low levels of greenhouse gas emissions (or none at all) compared to traditional vehicles based on fossil fuels [115–118]. However, PHEVs will significantly increase the load on the power grid. For example, if 30% of conventional vehicles in the United States were replaced by PHEVs, the total charging load would be around 18% of the peak US summer peak load [13, 14]. Fortunately, the traditional grid is transforming into a *smart grid*, which tends to be more efficient at generating

Smart Grid Communication Infrastructures: Big Data, Cloud Computing, and Security,
First Edition. Feng Ye, Yi Qian, and Rose Qingyang Hu.
© 2018 John Wiley & Sons Ltd. Published 2018 by John Wiley & Sons Ltd.

and transferring electricity to the customers. Because of advanced two-way communication networks, demand-side management (DSM) can be achieved by exchanging the information of customer-side load request and the corresponding control message (e.g. smart price or direct scheduling) [14, 100, 101, 105]. By applying DSM, the load of the power grid can be smoothed. In particular, many studies have been conducted to reduce the peak-to-average ratio (PAR).

In this chapter, we introduce PHEVs into the DSM system to further smooth the load of the power grid. Instead of letting customers charge on their own [14, 115, 119–122], it has been proposed to establish battery exchange stations (BESs) analogous to gas stations if PHEVs have standardized batteries [123]. In a BES, the charging outlets can be upgraded to more powerful ones (e.g. 600 volt [115]), which can charge the batteries much faster than normal power outlets (e.g. 120 volt in the United States) in households. Customers do not need to worry about spending a long time charging their PHEVs. It seems clear that if the BESs sell electricity back to the grid during peak hours and charge the batteries during off-peak hours, the load on the grid can be smoothed. Therefore, our focus for the proposed DSM system is on the BESs instead of individual PHEVs.

We first assume that the control center takes over the load scheduling of all the BESs, with the object of achieving minimum PAR while satisfying all the customers. This scheduling scheme is formed as a convex optimization problem where the optimal solution can be found. However, direct load control is not practical in the smart grid, so many existing DSM systems apply a *smart pricing* scheme. When a smart pricing scheme is used, electricity is more expensive when the load on the grid is high. Because of the real-time information exchange between customers and the control center, smart pricing can be applied frequently and regularly (e.g. hourly). Based on smart pricing, we propose another incentive load scheduling scheme for BESs so that they will make a profit by selling energy back to the grid during peak hours and charge the batteries during off-peak hours. With reasonable assumptions, we prove that this scheme also achieves minimum PAR. Moreover, we propose a game theoretical scheme so that the load scheduling can be done locally at each BES with little information exchanged. In this scheme, each BES tends to maximize its own profits instead of maximizing overall profits of all BESs. Unlike the first two schemes, the distributed scheme does not guarantee 100% customer satisfaction, since each BES lacks accurate information about the others. We conduct analysis on the planning of BESs (e.g. number of charging ports and amount of battery storage) in order to decrease the impact on customers.

7.1.2 Related Work

The PHEV charging scheduling problem has been studied by many researchers [14, 92, 115, 120–123]. Huang et al. in [92] studied the charging schedule for PHEVs in shared parking lots. However, while the peak load is reduced, the control center took over all the scheduling, which may not be practical. Fan in [14] applied a smart pricing scheme to distributed PHEV charging scheduling, in which each PHEV adjusts the demand according to the result of demand response. However, selling electricity back to the grid was not considered. Bahrami et al. in [115] also introduced smart pricing and formulated a game among all the users. The goal in their work was to minimize the cost of charging for all the owners. However, the users did not actively contribute to the DSM. Zhou et al. in [124] proposed distributed charging scheduling for PHEVs, which could avoid bus congestion and large voltage drops in the distribution grid. The PHEVs were provided with incentives, which is different from smart pricing. Moreover, the ability to sell back capability and DSM were not studied.

The authors in [120–122] applied a smart pricing scheme to encourage the PHEVs to sell back energy during peak hours. Moein et al. in [122] proposed a two-stage charging control strategy in a multicarrier energy environment. The customers were motivated to participate. However, the customers would passively adjust the load according to the smart pricing (time-of-use pricing in [122]). Liu et al. in [121] assessed the impact of the scheduling of PHEV charging on wind power; however, the DSM system was not considered. Dong et al. in [123] proposed adaptive scheduling of PHEV charging based on a battery replacement strategy (similar to BES). Their focus was on the impact of imperfect data communication on the scheduling performance; however, the BESs were not considered to contribute to the DSM system. In this chapter, we focus on BESs instead of direct charging of PHEVs. By introducing a smart pricing strategy, the BESs contribute positively to the DSM system by selling back electricity during peak hours, which will reduce the PAR of the power grid. The batteries stored in BESs form a power storage system; however, it is different from the traditional ones [125, 126]. While the traditional power storage system is designed to save overgenerated energy for future use, the BESs are intended to fulfill the needs of the PHEVs while lowering the PAR for the power grid. More specifically, energy stored in BESs does not decrease the requirement of power generation or increase the available energy when needed.

The rest of this chapter is organized as follows. In section 7.2, we illustrate the DSM being studied. In section 7.3, we propose several load-scheduling schemes for BESs. In section 7.4, we show the numerical and simulation results. In section 7.5, we present our conclusion and describe some future research directions.

7.2 System Model

In this section, the system model is illustrated. Notations, variables, and preliminary assumptions to be used in the problem formulation are also described in this section.

7.2.1 Overview of the Studied System

The DSM system is illustrated in Figure 7.1. It consists of a control center, power suppliers, and customers. Although DSM should be applied to all types of appliances, including residential, business, and industrial customers as well as PHEVs, we focus on the role of PHEVs in the DSM system in this section.

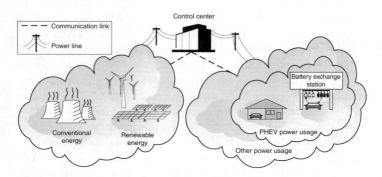

Figure 7.1 Demand-side power management system with BESs.

7.2.2 Mathematical Formulation

For better understanding of in problem formulation, all notations and variables to be used in the rest of this chapter are listed in Table 7.1.

Let $S = \{s_1, s_2, \ldots, s_N\}$ be the set of all BESs. The total number of BESs is assumed to be $N \triangleq |S|$. Analogous to traditional gas stations, BESs are assumed to be widely available throughout a town or city. Each BES is equipped with m_i fast charging outlets (e.g. DC fast charging, 600 volt) and discharging outlets. In practice, an outlet may be connected to multiple charging ports. We assume that only one port is connected to an outlet for simplicity. A BES does not directly provide charging/discharging ports for customers in order to maximize the utility of the equipment. Moreover, customers would only need to stop for a few minutes to exchange their batteries instead of for hours if charging them. Let $\mathcal{A}$ be the set of all PHEVs, and let $A \triangleq |\mathcal{A}|$ be the

Table 7.1 List of notations and variables.

Sets	
S	set of BESs $\{s_1, s_2, \ldots, s_N\}$
$\mathcal{T}$	set of time intervals $\{t_1, t_2, \ldots, t_T\}$
$\mathcal{A}$	set of PHEVs $\{a_1, a_2, \ldots, a_A\}$
$\mathbf{l}_i$	set of load scheduling for $s_i, \{l_i(t) | t \in \mathcal{T}\}$

Variables and parameters	
m_i	number of power outlets at s_i
B_i	number of batteries at s_i
$l_i^+(t)$	electricity bought from the grid at s_i during t
$l_i^-(t)$	electricity sold to the grid at s_i during t
$b_i^+(t)$	number of fully charged batteries at s_i during t
$n_i^+(t)$	number of customers at s_i during t
l_{max}^+	maximum charging electricity load for one outlet
l_{max}^-	maximum discharging electricity load for one outlet
l_0	amount of electricity to charge/discharge a battery
p_0	unit price for battery exchange
$p(t)$	unit price for electricity at t
τ_0	minimum charging time for a battery

total number of PHEVs. Although the DSM system can be modeled for any arbitrary time period to satisfy the assumptions, we consider a daily model in this work without loss of generality. Let each day be divided into several uniform time intervals (e.g. hourly), denoted as $\mathcal{T} = \{T_1, T_2, \ldots, t_T\}$.

We assume that the PHEVs are equipped with standardized batteries that can be easily exchanged. Without loss of generality, we assume that there is only one type of standard battery. It is relatively straightforward to extend the model with multiple sizes and types of batteries depending on the types of vehicles, similar to grades of gasoline. When a customer comes, a simple battery exchange is performed at a *flat rate* p_0 per exchange. For simplicity, we assume that the state of charge (SoC) of the batteries being exchanged is a constant value (e.g. 15%). A customer will exchange a 15% battery for a fully charged one. If the BES sells back the electricity to the grid, it discharges a fully charged battery to 15% as well. Besides earning profits by exchanging batteries, each BES also makes profits by participating in the DSM system with smart pricing mechanism. Each BES is willing to sell back electricity to the power grid during peak hours, when the price is much higher than during off-peak hours, and to charge the batteries during off-peak hours. Many researchers have considered the smart pricing function as convex and increasing [27–29]. It is generally adopted as a quadratic function as follows:

$$c(l) = al^2 + bl + c, \tag{7.1}$$

where a, b, c should be determined by utility companies. Note that l in Eq. (7.1) is the total load of the power grid but not the total load for PHEVs.

7.2.3 Customer Estimation

It is important to estimate the number of customers and their expected usage patterns for a BES business. If there is a large number of customers, a BES will require more equipment. If most customers visit during the same hour, then more batteries shall need to be prepared beforehand. More ports will be required if batteries are charged frequently. Without a good estimate, a BES may not be able to maximize its profits due to insufficient equipment or may even lose money due to deploying too much equipment. Moreover, efficient operation of BESs is required for the smart grid to achieve the best DSM.

Without loss of generality, we assume that most customers either go to a BES on their way to work or on their way back home. Each PHEV needs to exchange its battery (or charge) once daily. Moreover, a customer will not purchase service at BES s_i if no fully charged batteries are available. This customer will check some other BESs within the area during the same time slot (e.g. one hour). Therefore, it is assumed that the distribution of incoming customers at BESs follows the pattern of the customers' departure from and arrival at home. Following [124, 127, 128], the departure and arrival times of the PHEVs are modeled as two normal distributions with the mean of 7 a.m. and 6 p.m. and a standard deviation of one hour respectively. The distributions are expressed as follows:

$$p_d = \frac{1}{\sigma\sqrt{2\pi}} e^{-\frac{(t-\mu_d)^2}{2\sigma^2}}, \qquad (7.2)$$

and

$$p_a = \frac{1}{\sigma\sqrt{2\pi}} e^{-\frac{(t-\mu_a)^2}{2\sigma^2}}, \qquad (7.3)$$

where $\sigma = 1$, $\mu_d = 7$ and $\mu_a = 18$. Since a customer only needs to exchange the battery once, the estimate of incoming customers at BESs adopts modified distributions, that is, $0.5p_d$ for the customers coming during the departure window and $0.5p_a$ for the customers who come during the arrival window. The distribution of incoming customers to all the BESs is illustrated in Figure 7.2. Note that the estimate of incoming customers could be extended to more precise models based on big data analytics on the lifestyles of drivers and types of PHEVs from voluntary participants: for example, docked PHEVs at BESs, social networks, or tracking applications in smart phones. Moreover, we agree that not all customers will need to replace batteries on a daily basis in practice. Nonetheless, the service frequency of each customer can be easily extended to other settings.

In this chapter, assuming that each BES exchanges only one battery for each customer. Customers are not allowed to purchase extra batteries. The total number of batteries in a BES (i.e. s_i) is a constant value, B_i. Let τ_0 (e.g. hour) be the minimum charging time for a battery; then the maximum number of batteries can be charged daily at s_i is $\lfloor \frac{24m_i}{\tau_0} \rfloor$. Therefore, a larger B_i may cause waste instead of providing better service to customers or making more contribution to the DSM system.

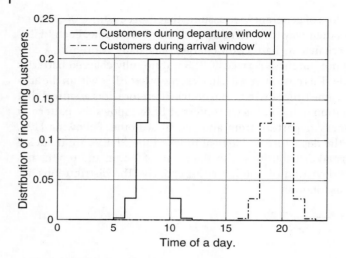

Figure 7.2 Distribution of incoming customers.

7.3 Load Scheduling Schemes for BESs

In this section, we propose centralized optimization schemes to minimize PAR. The minimum PAR can be used as a benchmark for distributed schemes or even real-time schemes based on dynamic settings. We also propose a game theoretical approach for practical deployment.

7.3.1 Constraints for a BES s_i

Let $l_i(t) = l_i^+(t) - l_i^-(t)$ be the electricity load at BES s_i during time slot t, where $l_i^+(t)$ is the load selling back to the grid, $l_i^-(t)$ is the load buying from the grid, and $n_i(t)$ is the number of cars exchanging batteries. Note that $l_i(t)$ can be negative, meaning that the BES is providing electricity to the power grid. Without loss of generality, we assume that the total load of the grid is always positive with DSM. We then discuss in detail the constraints for BES s_i.

During time slot t_i, the total amount of electricity bought from the grid cannot exceed the charging capacity of the station, that is

$$l_i^+(t) \leq l_{max}^+ \cdot m_i, \tag{7.4}$$

where $l_{\max}^+$ is the maximum charging electricity load for one port. Similarly, the total amount of electricity sold back to the grid cannot exceed the discharging capacity of the station, that is

$$l_i^-(t) \leq l_{\max}^- \cdot m_i, \tag{7.5}$$

where $l_{\max}^-$ is the maximum discharging electricity load for one port. Note that s_i does not buy and sell electricity during the same time slot, thus,

$$l_i^-(t) \cdot l_i^+(t) = 0, \ \forall t \in \mathcal{T}. \tag{7.6}$$

However, since s_i has limited storage space for fully charged batteries $b_i(t)$ during t_i, the total amount of electricity sold back to the grid is also bounded by $b_i(t)$, that is

$$l_i^-(t) \leq l_0 \cdot b_i(t), \ \forall t \in \mathcal{T}. \tag{7.7}$$

It is possible that s_i does not have enough fully charged batteries $b_i(t)$ for $n_i(t)$ customers. In this case, we assume that unsatisfied customers will randomly find another nearby station (e.g. s_j) with available batteries during the same time slot. Therefore, we must ensure that the overall number of fully charged batteries is enough for all the customers, that is

$$\sum_{i=1}^{N} b_i(t) \geq \sum_{i=1}^{N} n_i(t), \ \forall t \in \mathcal{T}. \tag{7.8}$$

A battery being charged will not be able to sell its energy back to the power grid. The electricity sold back to the grid must be from fully charged batteries. Batteries changed to the customers' PHEVs are also fully charged ones. In order to satisfy all the customers while reducing peak load on the power grid, the constraint that needs to be satisfied is as follows:

$$\sum_{i=1}^{N} l_i^-(t) + \sum_{i=1}^{N} (l_0 \cdot n_i(t)) \leq \sum_{i=1}^{N} (l_0 \cdot b_i(t)), \ \forall t \in \mathcal{T}. \tag{7.9}$$

Finally, the total surplus electricity bought from the grid should exceed the total need of the PHEVs, that is

$$\sum_{i=1}^{N} \sum_{t=1}^{T} l_i(t) \geq \sum_{i=1}^{N} \sum_{t=1}^{T} (l_0 \cdot n_i(t)) = l_0 \cdot A. \tag{7.10}$$

Note that the right hand side of Eq. (7.10) is a constant. A strict equality applies to this constraint if the system has 100% efficiency.

7.3.2 Minimizing PAR: Problem Formulation and Analysis

The objective of applying the DSM system in BESs is to smooth the PAR in the smart grid. Therefore, we first propose $\mathcal{P}1$ to directly minimize PAR, that is

$$\mathcal{P}1 : \quad \min \frac{\sup_{t \in \mathcal{T}} \left(\sum_{i=1}^{N} l_i(t) \right)}{\frac{1}{N} \sum_{i=1}^{N} \sum_{t=1}^{T} l_i(t)} \tag{7.11}$$

subject to *Eq.* (7.4), *Eq.* (7.5), *Eq.* (7.6), *Eq.* (7.7),
Eq. (7.8), *Eq.* (7.9), *Eq.* (7.10), $\forall s_i \in S$.

Lemma 7.1 If the total power load $\mathbf{L} = \{ \sum_{i=1}^{N} l_i(t) | \forall t \in \mathcal{T} \}$, then $\mathcal{P}1$ has a unique optimum solution $\mathbf{L}_1^{\star}$.

Proof: First, since $\frac{1}{N} \sum_{i=1}^{N} \sum_{t=1}^{T} l_i(t) = \frac{l_0 \cdot A}{N}$ is a constant value, the objective function can therefore be represented as follows:

$$\mathcal{P}1 \triangleq \min_{s_i \in S} \sup \left(\sum_{t=1}^{T} l_i(t) \right),$$

because $\sup_{s_i \in S} \left(\sum_{t=1}^{T} l_i(t) \right)$ is a convex function and the constraint set is compact, convex, and nonempty. Therefore, $\mathcal{P}1$ has a unique solution as follows:

$$\mathbf{L}_1^{\star} = \arg \min \left(\sup_{t \in \mathcal{T}} \sum_{i=1}^{N} l_i(t) \right). \tag{7.12}$$

$\square$

7.3.3 Problem Formulation and Analysis for Minimizing Costs

If all the customers choose to exchange their batteries at BESs instead of charging on their own, all of the batteries must be scheduled by the BESs. In this case, the optimal solution to $\mathcal{P}1$ is the minimum PAR. When some customers choose to charge PHEVs on their own, those charging loads will not purposely contribute to lowering the PAR. Thus the overall PAR will be higher than the optimal PAR in most cases. However, we still need to make it clear that both the BESs and the customers have the incentive to participate in the DSM system, because the BESs can make more profits while customers can generate more savings. Let $p(t)$ be the unit cost of electricity (i.e. $/KWh$), which follows the cost function shown in Eq. (7.1). Let the daily profit of BES s_i be the utility, that is

$$p(t) = c \left(\sum_{i=1}^{N} l_i(t) + L(t) \right), \tag{7.13}$$

where $L(t)$ is the total load of other appliances in the power grid after applying DSM. Note that $L(t)$ is considered as a constant in this work.

An intuitive incentive for each station is the overall daily profit, which is calculated as follows:

$$u_i = - \sum_{t=1}^{T} (p(t) \cdot l_i(t)) + \sum_{t=1}^{T} p_0 \cdot n_i(t). \tag{7.14}$$

Note that s_i may not satisfy $n_i(t)$ on its own; we assume that all the BESs together satisfy all the customers in this centralized scheme. The objective is to maximize the overall profit of the BESs, that is

$$\mathcal{P}2: \quad \max \sum_{i=1}^{N} u_i, \tag{7.15}$$

subject to *Eq. (7.4), Eq. (7.5), Eq. (7.6), Eq. (7.7),*
Eq. (7.8), Eq. (7.9), Eq. (7.10), $\forall s_i \in S$.

Lemma 7.2 $\mathcal{P}2$ has a unique optimum solution $\mathbf{L}_2^{\star}$.

Proof: Note that $\sum_{i=1}^{N} \sum_{t=1}^{T} p_0 \cdot n_i(t)$ is a constant; therefore, problem $\mathcal{P}2$ can be represented as follows:

$$\mathcal{P}2 \triangleq \min \sum_{i=1}^{N} \sum_{t=1}^{T} (p(t) \cdot l_i(t)).$$

Since $\sum_{i=1}^{N} \sum_{t=1}^{T} l_i(t) = l_0 \cdot A$ is a constant and $p(\cdot)$ is a strict convex function with respect to t, therefore $\sum_{i=1}^{N} \sum_{t=1}^{T} (p(t) \cdot l_i(t)) = \sum_{t=1}^{T} (p(t) \sum_{i=1}^{N} l_i(t))$ is a convex function. With the same compact, convex, and nonempty constraint set, $\mathcal{P}2$ has a unique solution as follows:

$$\mathbf{L}_2^{\star} = \arg \min \sum_{t=1}^{T} \left(p(t) \sum_{i=1}^{N} l_i(t) \right). \tag{7.16}$$

□

Theorem 7.1 $P1$ and $P2$ have the same optimum solution with respect to the total load schedule $\mathbf{L} = \{\sum_{i=1}^{N} l_i(t) | t \in \mathcal{T}\}$.

Proof: Let $\mathbf{L}_1^\star \triangleq \mathbf{x} = \{x_t | t \in \mathcal{T}\}$ be the optimal solution of $P1$, and let $\mathbf{L}_2^\star \triangleq \mathbf{y} = \{y_t | t \in \mathcal{T}\}$ be the optimal solution of $P2$. Without loss of generality, we reorganize $\mathbf{x}, \mathbf{y}$ to be nondescending sets such that $x_i \leq x_{i+1}$, $y_i \leq y_{i+1}$, $i \in \mathcal{T}$ for a clearer illustration. Note that $\max_{t \in \mathcal{T}} \mathbf{x} = x_{|\mathcal{T}|}$, and $\max_{t \in \mathcal{T}} \mathbf{y} = y_{|\mathcal{T}|}$. Assuming $\mathbf{x} \neq \mathbf{y}$, then we have

$$x_T < y_T, \tag{7.17}$$

$$\sum_{i \in \mathcal{T}} x_i p(x_i) > \sum_{i \in \mathcal{T}} y_i p(y_i). \tag{7.18}$$

Furthermore, inequality (7.18) indicates that

$$\sum_{i \in \mathcal{T} \setminus \{T\}} (x_i p(x_i) - y_i p(y_i)) > y_T p(y_T) - x_T p(x_T). \tag{7.19}$$

The left-hand side of inequality (7.19) is maximized when $x_i = y_i = 0$, $i \in \mathcal{T} \setminus \{\mathcal{T}|\}$ since $\sum_{i=1}^{N} x_i = \sum_{i=1}^{N} y_i = l_0 \cdot A$ is a constant, and function $g(x) = x p(x)$ is increasing and strict convex with respect to x. Therefore, we have

$$(l_0 \cdot A - x_T) p(l_0 \cdot A - x_T) - (l_0 \cdot A - y_T) p(l_0 \cdot A - y_T)$$

$$\geq \max_{\mathbf{x}, \mathbf{y} \in \mathbb{L}} \left(\sum_{i \in \mathcal{T} \setminus \{T\}} (x_i p(x_i) - y_i p(y_i)) \right). \tag{7.20}$$

However, because of inequality (7.17), then

$$x_{T-1} \leq l_0 \cdot A - x_T < x_T,$$

$$y_{T-1} \leq l_0 \cdot A - y_T < y_T. \tag{7.21}$$

and the fact that

$$l_0 \cdot A - x_T - l_0 \cdot A - y_T = y_T - x_T. \tag{7.22}$$

Therefore, we can see that

$$\sup_{\mathbf{x}, \mathbf{y} \in \mathcal{L}} \sum_{i \in \mathcal{T} \setminus \{T\}} (x_i p(x_i) - y_i p(y_i))$$

$$\leq (l_0 \cdot A - x_T) p(l_0 \cdot A - x_T) - (l_0 \cdot A - y_T) p(l_0 \cdot A - y_T)$$

$$\leq y_T p(y_T) - x_T p(x_T). \tag{7.23}$$

Note that inequality (7.23) contradicts inequality (7.19); thus $\mathbf{x} = \mathbf{y}$ (or $\mathbf{L}_1^\star = \mathbf{L}_2^\star = \mathbf{L}^\star$ with proper ordering), which completes the proof. $\quad\square$

Existing convex optimization problem solutions can be applied to reach the optimal solution $L_\star$.

Other Issues to Be Considered In $P2$, we exclude $\sum_{t=1}^{T} p_0 n_i(t)$ from the objective functions since it returns a constant value. However, p_0 must be determined so that the customers are willing to exchange their batteries instead of charging on their own. After solving $P1$ or $P2$, we find the optimal load schedule $L^\star$, then we need to determine p_0 so that the customers will spend more to charge on their own, that is

$$p_0 \leq \sum_{t_s}^{t_e} \left(p(t) \cdot \frac{l_0}{t_e - t_s} \right), \tag{7.24}$$

where t_s is the time to start charging and t_e is the time to end charging. Moreover, since a household does not have as powerful an outlet as BESs do, customers will prefer a battery exchange if the cost is reasonable. Moreover, we need to determine the number of batteries B_i carried by each BES s_i so that the number of units in storage will be enough while unnecessary storage can be avoided.

7.3.4 Game Theoretical Approach

In practice, it is impossible to let a control center schedule all the BESs. Technically, a centralized system is too complex to achieve the computational efficiency required by the scale of the smart grid. Besides, very few BESs would surrender their confidential information to the utility company. Therefore, distributed local scheduling schemes are more appropriate for the smart grid. For this reason, we propose a game theoretical approach for the DSM system. The game is formulated as a 3-tuple $G = (S, \{\mathcal{L}_i\}, \{u_i(\cdot)\})$, where $\mathcal{L}_i$ is the set of all possible load schedules for station i. Let l_{-i} be the load scheduling of all the BESs $s_j \in S \backslash s_i$. The utility for each BES is the daily profit, represented as follows:

$$u_i(l_i, l_{-i}) = - \sum_{t=1}^{T} (p(t) \cdot l_i(t)) + \sum_{t=1}^{T} (p_0 \cdot [b_i(t), n_i(t)]^-), \tag{7.25}$$

where $[\cdot]^-$ returns the smaller value of the two inputs.

In the proposed local scheduling scheme, each BES (e.g. s_i) is limited by the physical charging/discharging capacity, as expressed in Eq. (7.4), Eq. (7.5), and Eq. (7.6). Note that with a local scheduling scheme, if a customer cannot exchange a battery at s_i, there is no guarantee that another BES will be available. Although s_i does not guarantee 100% customer satisfaction, its storage of fully charged batteries satisfies some of the customers and also allows electricity to be sold back to the grid, that is

$$l_i^-(t) + l_0 \cdot [b_i(t), n_i(t)]^- \leq l_0 \cdot b_i(t), \ \forall t \in \mathcal{T}. \tag{7.26}$$

For s_i, the total surplus electricity bought from the grid should exceed the total need of the PHEVs at s_i, that is

$$\sum_{t=1}^{T} l_i(t) \geq \sum_{t=1}^{T} (l_0 \cdot [b_i(t), n_i(t)]^-). \tag{7.27}$$

The objective for each BES s_i is to maximize its daily profit with the given smart pricing schedule, that is

$$\mathcal{P}3 : \ \max u_i. \tag{7.28}$$

subject to *Eq.* (7.4), *Eq.* (7.5), *Eq.* (7.6), *Eq.* (7.7),

Eq. (7.26), *Eq.* (7.27).

Note that the constraints are applied to individual BES s_i only. For simplicity, we assume that $n_i(t)$ has taken into consideration the customers not served by another BES. Therefore, a BES should always have enough fully charged batteries for incoming customers. In this case, we have $b_i(t) \geq n_i(t), \ \forall s_i \in S, \forall t \in \mathcal{T}$. If a customer comes in without being serviced, he/she will leave and charge on his/her own at a public parking lot or at home.

Lemma 7.3 With given $p(t)$ and $\mathbf{l}_{-i}$, $\mathcal{P}3$ has a unique optimal solution $\mathbf{l}_i^\star$.

Proof: $\mathcal{P}3 \triangleq \min \sum_{t=1}^{T} (p(t) \cdot l_i(t))$, where the objective function is strictly convex. With compact, convex, and nonempty constraint set, $\mathcal{P}3$ has a unique solution $\mathbf{l}_i^\star = \arg \min \sum_{t=1}^{T} (p(t) \cdot l_i(t))$. □

Lemma 7.4 An NE $\mathbf{l}^\star$ uniquely exists for $\mathcal{G}$.

Proof: We show the sketch of the proof. First, the payoff function (Eq. 7.25) is strictly concave, and the constraint set for this approach is

---

Algorithm 7.1 Algorithm to find NE of $\mathcal{G}$

Input: $n_i(t)$, $\forall s_i \in S$, $\forall t \in \mathcal{T}$;
Output: $\mathbf{l}_i^\star$, $\forall s_i \in S$;
1: Each player computes a feasible $\mathbf{l}_i$
2: **while** NE is not achieved **do**
3: $\quad \{p(t) | t \in \mathcal{T}\} \leftarrow p(t) = c\left(\sum_{i=1}^{N} \mathbf{l}_i\right)$; // Computed and distributed
$\quad$ by the control center
4: $\quad \mathbf{l}_i \leftarrow \mathbf{l}_i^\star = \arg \mathcal{P}3$, $\forall s_i \in S$; // Computed by s_i
5: **end while**
6: $\mathbf{l}_i^\star \leftarrow \mathbf{l}_i$, $\forall s_i \in S$; // Output NE

---

compact, convex, and nonempty; thus NE exists. Second, lemma 7.3 guarantees that the best response of each player can be found uniquely. Therefore, NE exists uniquely to $\mathcal{G}$. $\qquad\qquad\square$

Definition 7.1 *Nash equilibrium (NE)* A scheduling set $\mathbf{l}^\star = (\mathbf{l}_1^\star, \ldots, \mathbf{l}_N^\star)$ is an NE of $\mathcal{G} = (\mathcal{N}, \{\mathcal{L}_i\}, \{u_i(\cdot)\})$ if, $\forall s_i \in S$, $\forall \mathbf{l}_i \in \mathcal{L}_i$, $u_i(\mathbf{l}_i^\star, \mathbf{l}_{-i}^\star) \geq u_i(\mathbf{l}_i, \mathbf{l}_{-i}^\star)$, where $\mathbf{l}_{-i}^\star = (\mathbf{l}_1^\star, \ldots, \mathbf{l}_{i-1}^\star, \mathbf{l}_{i+1}^\star, \ldots, \mathbf{l}_N^\star)$.

The NE can be found by iteratively calculating the best response at each BES following Algorithm 7.1. In each iteration, each BES submits its load schedule to the control center, then the control center computes the price schedule and sends it back to the BESs. The BESs will adjust their schedules based on the given price schedule and resubmit them to the control center.

7.4 Simulation Analysis and Results

In this section, we conduct a simulation to demonstrate the proposed intelligent charging schemes for BESs in the smart grid.

7.4.1 Settings for the Simulations

In the simulation, we assume that there are 10 BESs in the town. For simplicity, the BESs are assumed to be identical. A total of $A = 500$ PHEVs require battery exchange service daily in this area. Each battery is assumed to have a capacity of 10 kilowatt-hours, thus $l_0 = 8.5$ kilowatt-hours as 85% of the capacity. Without loss of generality, we assume that each charging port at the BES charges a battery at 17 KW per hour with 100% efficiency and discharges a battery at 8.5 KW per hour with 100% efficiency. Let $T = 24$ so that each time slot lasts one hour. In this case, two batteries can be fully charged using the same port in one time slot, and one battery can be depleted in one time slot. These settings will be used throughout the simulations unless specified otherwise. Note that the settings are chosen arbitrarily for performance evaluation of the proposed DSM. The settings are subject to change based on further big data analytics before actual deployment of BESs.

To better demonstrate the impact of PHEVs on the smart grid, we assume the existing load of the power grid without PHEVs follows a trend of typical peak and off-peak hours. The off-peak hours span from 0 to 8 and from 19 to 24, while peak hours span from 9 to 18. An example of the power load in a power grid without PHEVs is shown in Figure 7.3. Peak and off-peak hours can be seen in the illustration. Note that the exact power load in a time slot fluctuates. The total load needed by PHEVs ($l_0 \cdot A$) is assumed to be exactly 18% of the existing load in the power grid for better evaluation of the proposed schemes.

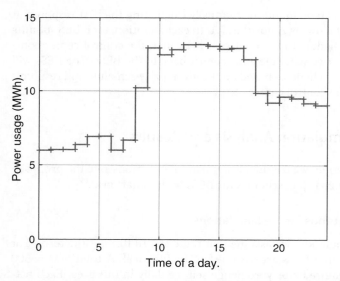

Figure 7.3 An example of a power load without PHEVs in the power grid.

7.4.2 Impact of the Proposed DSM on PAR

In this evaluation, we show that intelligent charging schemes for PHEVs can contribute to smooth PAR in the smart grid. Assume that each BES is equipped with an unlimited supply of batteries, 20 charging ports, and $b_i(1) = 20$ fully charged batteries in the beginning. The solution to the PAR minimization problem $\mathcal{P}1$ can be found by solving Eq. (7.16). The optimal load schedule is shown in Figure 7.4. We can see that with the introduction of BESs, although the total load of the power grid increases by 18%, the peak load is reduced. As a result, the load schedule is further smoothed, and the PAR is reduced. Note that because the BESs sell back electricity to the grid during previous peak hours and charge during previous off-peak hours, the new load schedule has a different pattern of peak/off-peak hours. For example, previous off-peak hours from midnight to 8 a.m. are peak hours with DSM. Therefore, the concept of peak/off-peak hours may be dropped in the smart grid. Instead, real-time based smarting pricing may be applied, as discussed in the previous chapter.

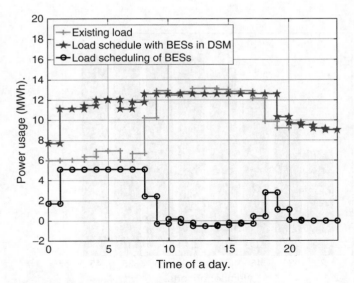

Figure 7.4 Smoothed load with BESs.

7.4.3 Evaluation of BESs Equipment Settings

7.4.3.1 Number of Charging Ports

The numbers of charging/discharging ports and batteries have different impacts on BESs as well as the efficiency of the DSM. For example, increasing the number of charging ports would increase the parallel charging ability of the BESs. A larger number of batteries would increase the service capacity of a BES as well its contribution to the DSM system. However, the number of batteries is limited by electricity demand and electricity supply in the smart grid. With the settings applied in this simulation, the number of stored batteries reaches an upper bound as the number of ports increases to support the DSM most efficiently, as shown in Figure 7.5. Therefore, there is no need to deploy too many charging ports at each BES. Such information should be provided by utility companies to avoid shortages or overdeployment.

7.4.3.2 Maximum Number of Fully Charged Batteries

Another important parameter to each BES is the number of fully charged batteries available in a given time slot. Intuitively, more

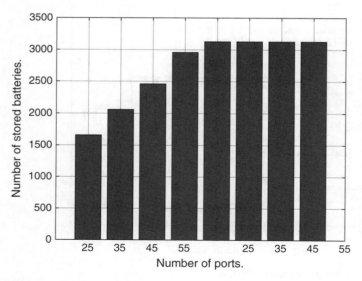

Figure 7.5 The impact of m_i on fully charged batteries.

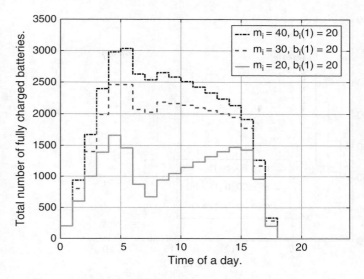

Figure 7.6 Impact of port number on fully charged batteries.

of such batteries would allow the BES to provide service to more customers and contribute more to the DSM system. However, storing more fully charged batteries requires more ports for charging. With the same settings used in the simulation before, the minimum number of fully charged batteries required in each time slot is shown in Figure 7.6. In this simulation, the maximum number of fully charged batteries is required at six a.m. The BESs take advantage of the previous off-peak hours during the night to prepare for the service peak to incoming customers as well as to resell electricity. Therefore, the optimal number of fully charged batteries stored at each BES shall be the maximum requirement of each time slot. In the case evaluated in this part, each BES will store $\sum_{i=1}^{N} B_i(6)$ fully charged batteries to support PHEV customers and DSM optimally. In reality, this value could be used as a benchmark for BESs, based on their actual service capacity.

7.4.3.3 Preparation at the Beginning of Each Day

The number of fully charged batteries at the beginning of each day $(b_i(1))$ has an impact on the total amount of batteries as well. With $m_i = 30$ fixed, we show the impact of different $b_i(1)$ in Figure 7.7. As we can see, a larger $b_i(1)$ results in a larger amount of battery storage. Now

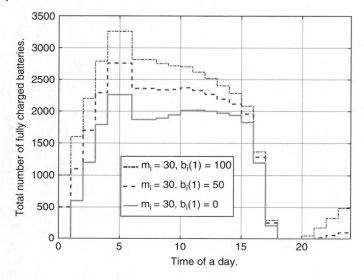

Figure 7.7 Impact of $b_i(1)$ on fully charged batteries.

that it is unlikely to sell back electricity during off-peak hours, it might be more cost effective for the BESs to start with $b_i(t) = 0.5Ap_d(1)/N$, where the storage is sufficient for incoming customers. Considering the estimates used in this chapter, $b_i(1) \approx 0$. However, whether $b_i(1) \approx 0$ is the optimal choice or not depends on PAR as well, since it affects the efficiency of the DSM system.

7.4.3.4 Impact on PAR from BESs
In this evaluation, we would like to present different contributions from BESs based on two factors: preparation for the beginning of each day and the number of ports. In Figure 7.8, we show the impact on PAR with different preparations for the beginning of each day (i.e. $b_i(1)$). It can be seen that a larger $b_i(1)$ leads to a lower PAR. This result occurs because hat once there are sufficient batteries for PHEV customers, the remaining batteries will be used as storage units to contribute to DSM. With a large enough number of storage units, the PAR can be reduced to 1 in the smart grid. However, as mentioned earlier, maintaining more storage units requires more charging/discharging ports in each BES.

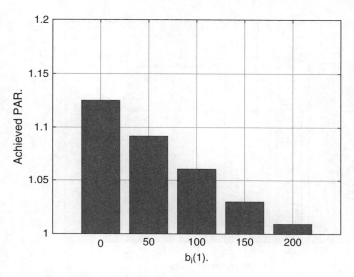

Figure 7.8 Impact of $b_i(1)$ on PAR.

The impact of the number of ports m_i on the PAR is shown in Figure 7.9. The results show that with more charging ports in each BES, a lower PAR could be achieved in the smart grid. With more charging ports, a BES can fully charge more batteries during off-peak hours and support the grid during peak hours. However, since the increase in battery storage has an upper bound, the achievable PAR has a lower bound. In this case, the optimal PAR (i.e. PAR=1) is not reached even with four times the number of ports. The relationship between the optimal number of charging ports and number of PHEVs needs to be studied in the future with big data analytics so that better guidance will be available for BESs deployment.

7.4.4 Evaluations of the Game Theoretical Approach

Using the general settings and $m_i = 20$, $b_i(t) = 20$, we evaluate the game theoretical approach. As shown in Figure 7.10, the distributed approach did not achieve the minimum PAR. Compared to the optimal solution achieved by solving problems $\mathcal{P}1$ or $\mathcal{P}2$, the game theoretical approach returns a higher peak demand with the 18% demand increase from PHEVs. Nonetheless, the PAR is lowered by almost 9% from 1.3158 to 1.2047, which clearly demonstrates that

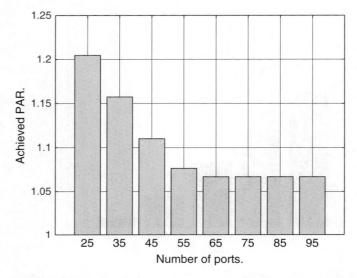

Figure 7.9 The impact of port number on PAR.

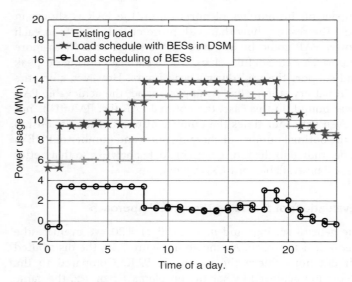

Figure 7.10 Load schedules achieved by distributed scheme.

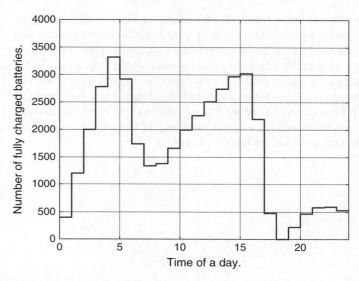

Figure 7.11 Estimate of battery storage by distributed scheme.

the proposed distributed approach could contribute to DSM in the smart grid. Correspondingly, the maximum number of fully charged batteries required in each time slot is given in Figure 7.11. Compared to the optimal solution, the distributed scheme requires more fully charged batteries.

7.5 Summary

In this chapter, we studied intelligent charging schemes for PHEVs in the smart grid. Instead of charging the PHEVs directly, we proposed to adopt battery exchange stations so that the customers only need to spend a few minutes exchanging their batteries. While PHEVs will significantly increase demand for electricity on the power grid, the intelligent charging schemes can be designed to cope with the demand increase and enhance grid efficiency. Because of the large number of batteries stored at each BES, we proposed to let the BESs participate in the demand-side management system so that they can store electricity during off-peak hours and reduce electricity demand from the grid during peak hours. Specifically, we proposed several load-scheduling schemes to minimize the PAR while earning extra profits for BESs. We successfully proved that the proposed profit incentive scheme also minimizes the PAR. We also proposed a distributed game theoretical approach so that the scheduling can be done locally in practice. In the simulations, we analyzed the optimal number of stored batteries and the selection of a reasonable number of charging ports to help building BESs. In future research, more types of PHEVs and versatile lifestyles should be considered for a better modeling of the demand-side management system based on big data analytics.

8

Big Data Analytics and Cloud Computing in the Smart Grid

The advanced communications infrastructure to be deployed in the smart grid is used for data exchange in many systems. Whether the smart grid can benefit from new and upgraded features, such as demand response and wide-area monitoring system is closely related to issues of collection, storage, and processing of data. In this chapter, we will introduce big data analytics and cloud computing and discuss their relevance to the smart grid. More specifically, we will focus on relevant issues in demand response and wide-area monitoring for illustration.

8.1 Background and Motivation

8.1.1 Big Data Era

Big data is a term for large and complex data sets that cannot be adequately processed with traditional data processing application software [129]. The concept of big data is widely defined as a set of 'V's:

- **Volume**. Big data indicates a massive volume of data that continues to grow larger.
- **Velocity**. Data comes in at unprecedented and possibly increasing speeds. Many data streams are generated in near real time.
- **Variety**. Data comes in all types of formats, structured and unstructured.
- **Variability**. The inconsistency of the data sets can hamper processes to handle and manage them.
- **Veracity**. Data quality varies greatly.

Smart Grid Communication Infrastructures: Big Data, Cloud Computing, and Security,
First Edition. Feng Ye, Yi Qian, and Rose Qingyang Hu.
© 2018 John Wiley & Sons Ltd. Published 2018 by John Wiley & Sons Ltd.

According to [130], the world's effective capacity to exchange information through telecommunication networks has increased from an annual volume of 281 petabytes in 1986 to an annual volume of 667 exabytes in 2014, as illustrated in Figure 8.1. The annual volume will continue to increase. The same research pointed out that general-purpose computing capacity grew at an annual rate of 58%. The world's capacity for bidirectional telecommunication grew at 28% per year, closely followed by the increase in globally stored information (23%).

Big data often refers to the use of predictive analytics, user behavior analytics, or other methods for data analysis that extract information from data sets. In particular, *big data analytics* would be a more appropriate term. Data analytics has been studied and applied for decades in areas such as machine learning and data mining. Compared to those traditional data analytics methods, big data poses challenges, including capture, storage, analysis, search, sharing, transfer, visualization, information privacy, etc. The issue of big data was initially introduced in space exploration, weather forecasting, medical genetic investigations, etc. Similar problems occurred to social media such as Facebook, Twitter, YouTube, and others. Big data is inevitable in the smart grid.

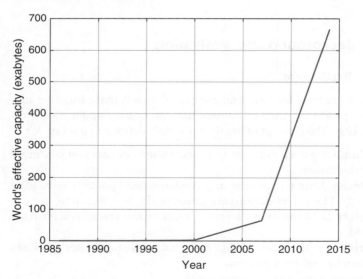

Figure 8.1 The world's effective data capacity.

8.1.2 The Smart Grid and Big Data

The smart grid is in the big data era because the presence of the 'V' concepts can be demonstrated. The smart grid has an advanced bidirectional communications infrastructure that enables many systems, such as advanced metering infrastructure (AMI) and wide-area monitoring systems (WAMS). It is foreseeable that smart grid will generate a massive amount of data. For instance, smart meters will soon generate over 1,000 petabytes a year in the United States [78]. Real-time data from the smart grid could be generated in terabytes due to the massive scale of the power grid. Thus the velocity of data is increasing considerably. Moreover, the smart grid has various data sources that range from field measurements obtained by substation/feeder intelligent electronic devices to specialized commercial and/or government/state databases, weather data of different types, lightning detection data, seismic data, fire detection data, electricity market data, vegetation and soil data, etc [131].

The processing and management of data on such a scale in the smart grid requires considering the data analytics used in big data industries. In general, data in the smart grid needs to be relevant, clean, organized, aggregated, processed, and analyzed to obtain and present actionable intelligence [132]. Cybersecurity and energy use are issues to be addressed with big data techniques, due to the massive number of devices in the communications infrastructure. The smart grid is intended to achieve efficient grid operations and to reduce greenhouse gas emissions. Therefore, the increases in energy consumption required by big data as well as cybersecurity should not threaten the operation efficiency of the grid.

In particular, demand response in the smart grid requires data analytics to accurately forecast electricity demand and predict peak power needs to power generators. Consumers rely on pricing/tariff set by demand response programs to control their electricity usage. Another application in the smart grid is its advanced monitoring and control system, for example WAMS. Transmission and distribution entities in the smart grid will use data analytics to identify anomalies in power delivery, detect and avert outages before they happen, and restore service faster after an outage. While data volume may not be *big* in the monitoring system, its real-time requirement presents challenges equivalent to those of big data.

8.2   Pricing and Energy Forecasts in Demand Response

Demand response is one of the most important applications in the smart grid that is based on the communication infrastructure (particularly the AMI). Although real-time demand response is the ultimate goal, online algorithms for controlling it are hard to deploy on a large scale. As an alternative, pricing forecasts and energy forecasts are critical to efficient demand response [133].

8.2.1   An Overview of Pricing and Energy Forecasts

A variety of data is gathered internally from the smart grid, including *internal data* and *external data*, as illustrated in Table 8.1. Among internal data, one critical set of data is the *metering data* generated by smart meters in the AMI. Metering data reveals the electricity usage of each household. Overall, metering data reveals the real-time (or nearly real-time) power usage of the grid from the customer side. Although utility companies already have monitored the electricity sold to the grid from the standpoint of power generators even in the traditional power grid, measurements of power usage from the customer side are more precise due to power losses during transmission and distribution [134, 135]. Other internal data (e.g. PMU data) is more important for monitoring transmission line status.

Apart from internal data generated in the smart grid, external data from other sources is another important input for optimizing control over the power grid. For example, the amount electricity to be generated from conventional power generators depends not only on energy requirements of customers but also on the capability of renewable resources and storage units. The capacity of renewable resources

Table 8.1 Useful data in the smart grid

Internal data	External data
Metering data	Weather forecast
PMU data	Social networks
Power line monitoring data	Stock markets
Etc.	Etc.

(e.g. a solar farm) is not likely to be controllable. Nonetheless, a precise weather forecast will be helpful for capacity prediction. External data can be different in types and sources. For example, it can come from weather forecasts, social networks, location-based tracking applications of smart devices, and many other sources.

The collected data lets the service provider optimize control over the power grid by giving the energy forecast to power generators and the price forecast to customers and smart appliances. As discussed before in chapter 6, it is critical to have schedules of appliances for demand response. With an appliance schedule determined, it is easy to have a schedule for power generation (including the electricity needed from renewable sources). As a consequence, the corresponding price can be found accordingly. Therefore, to provide price and energy forecasts, one good solution is to model the energy consumption of different appliances and (more importantly) the schedule of each appliance by the service provider. The energy consumption of each appliance may depend on its load; for example, an air conditioner (AC) uses more electricity if a lower temperature is set on its thermostat. However, the fluctuation of power is not too significant for many appliances (e.g. washing machines, coffee makers, etc.). Learning the schedules of customers can reveal much about the schedule of appliances. Without loss of generality, the following example residential customers. It has similar reasoning for other customers (i.e. industrial ones and business ones). To model the schedule of appliances in a house more precisely, one can extract useful information from the schedule of residential customers by getting answers to the following questions:

- When do they work?
- How far are they away from work?
- How do they commute?
- What do they do outside after work?
- What do they do at home?

With a model for the schedule of a customer, one may assume that the air conditioner (AC) is turned off (or set at a higher temperature) during working hours. The AC is turned on before the user returns home from work. The advance is based on the commuting time of the customer. The schedule is postponed for a few hours on Friday night if the customer schedules time for activities outside the house. The use of the AC remains steady throughout weekends. The usage patterns of many other appliances are closely related to activities of customers as well. In addition, the schedule and levels of energy consumption vary with changes in the surrounding environment, such as differences in temperature and humidity. The use of big data analytics in social networks is being widely studied [136]. A massive amount of useful information can be mined from the status posted by customers. For example, if a customer checks in at a hotel with his/her family (as seen via a post on Twitter or Facebook), then it is possible that their house will have low power usage in the coming days. A final energy forecast can be made after complex data analytics.

An overview of the data processing procedure is depicted in Figure 8.2. It includes three components: data input, big data analytics, and information output. Big data analytics is the core function of the process, which consists of four steps: 1) *data collection*, 2) *data preprocessing*, 3) *data storage*, and 4) *data analysis* [137].

8.2.2 A Case Study of Energy Forecasts

Machine learning is a tool for data analysis and predictions. In this section, a case study will be given to illustrate energy forecasting. The studied data set is a three-month record of "Home B" from [138]. It includes a data set of the energy consumption of "Home B" with a sampling frequency of 1 Hz, and a data set of the surrounding environmental conditions (e.g. inside temperature, outside temperature, humidity, etc.) with a sampling frequency of more than 10 times per hour. Detailed interpretation of the data sets can be found in [139].

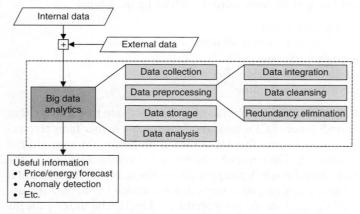

Figure 8.2 Data processing procedure.

In this case study, only daily consumption and average daily temperature are used.

Let $\mathbf{x} = (x_1, x_2, \dots)$ be the set of daily average temperatures (in Celsius). Let $\mathbf{y} = (y_1, y_2, \dots)$ be the set of daily energy consumption (in kilowatt-hours). Figure 8.3 depicts the energy consumption (the solid line) and the average temperature (the dashed line) of each day for a month.

Among all the appliances monitored, the air conditioner and heater have the highest levels of energy consumption. For simplicity, we assume that the daily energy consumption is closely related to the daily average temperature. Assume that y is some deterministic $f(x)$, plus random noise, such that $y = f(x) + \epsilon$, where $\epsilon \sim N(0, \sigma)$ follows a zero mean Gaussian distribution. Moreover, we assume that both $\mathbf{x}$ and $\mathbf{y}$ are independent and identically distributed random variables. Therefore, y is a random variable that follows the distribution

$$p(y|x) = N(f(x), \sigma). \tag{8.1}$$

Linear regression is applied as the learning algorithm; that is, $f(x) = w_0 + w_1 x$ is a linear function of x. Then we have

$$p(y|x) = N(w_0 + w_1 x, \sigma), \tag{8.2}$$

$$E[y|x] = w_0 + w_1 x. \tag{8.3}$$

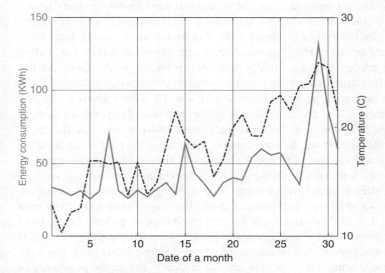

Figure 8.3 Energy consumption and temperature.

Alternatively, we have the likelihood as follows:

$$p(y|x; w_0, w_1) = N(w_0 + w_1 x, \sigma). \tag{8.4}$$

The optimal $< \hat{w}_0, \hat{w}_1 >$ is found by applying a maximum conditional likelihood estimate:

$$< \hat{w}_0, \hat{w}_1 > = \arg\max_{w_0, w_1} \prod_i p(y_i | x_i, w_0, w_1)$$

$$= \arg\max_{w_0, w_1} \sum_i \ln[p(y_i | x_i, w_0, w_1)], \tag{8.5}$$

where

$$p(y|x; w_0, w_1) = \frac{1}{\sqrt{2\pi\sigma^2}} \exp\left(-\frac{1}{2}\left(\frac{y_i - f(x_i; w_0, w_1)}{\sigma}\right)^2\right)$$

$$\tag{8.6}$$

Note that the logarithmic function can be applied because the variables are independent and identically distributed. After some straightforward arithmetic steps, we have

$$< \hat{w}_0, \hat{w}_1 > = \arg\min_{w_0, w_1} \sum_i (y_i - f(x_i; w_0, w_1))^2 \tag{8.7}$$

To solve Eq. (8.7), we can derive a gradient descent rule. Or we may apply heuristic optimization algorithms such as Monte Carlo and particle swarm optimization. The solution is used for energy forecasts.

In this case study, two types of energy forecasts are made (with results illustrated in Figure 8.4). The first type of energy forecast is made by using the training data of one month and then calculating the energy forecast for the next month based on the daily average temperature. The other type of the energy forecast is provided daily. For instance, the energy forecast for day 10 is only given at the end of day 9. The training data is gathered from the previous k days. Note that the real daily average temperature is used as the input of the prediction. It is reasonable to assume that a precise weather forecast can be obtained beforehand. The results of the monthly energy forecast are shown as the red solid line. As we can see, the monthly forecast gives a rough idea of energy consumption; however, it lacks accuracy. The results of the daily energy forecast are given with $k = 4, 7, 14$. We can clearly see that the daily energy forecast is closer to the real level of energy consumption. Although the energy forecasts are not precise due to the simplicity of the case study (only the daily average temperature is considered), it clearly shows the possibility of making precise energy forecasts with more factors considered and more advanced techniques for big data analytics.

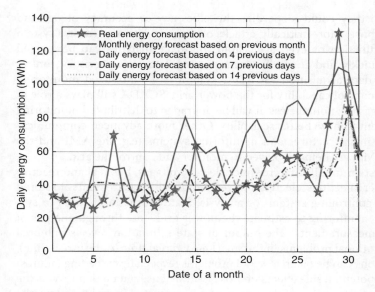

Figure 8.4 Energy forecast.

8.3 Attack Detection

8.3.1 An Overview of Attack Detection in the Smart Grid

Attacks on metering data may compromise the privacy of customers. Moreover, energy fraud can also be caused by attacks on metering data [140]. There are two types of energy fraud. In one, the customer reports less energy consumed than actually used. In the second, more energy may be used by rogue connections than is actually consumed by customers. Both types of the energy fraud will create deviations in optimal grid control, especially demand response. Specifically, if less energy is reported than is used, demand response will send a lower energy requirement to suppliers. That will in turn cause an insufficient electricity supply or a waste of fuel due to a sudden transfer from low to high levels electricity generation. If more energy usage is reported by rogue connections, the power station will overgenerate electricity, and that will increase the price unnecessarily.

Besides metering data, the control system in the smart grid also generates various monitoring data from different sensors, such as phasor measurement units (PMUs) or transmission line monitoring sensors. As mentioned above, if attacks are launched on metering data, they could jeopardize the efficiency or accuracy of demand response. More critically, attacks on sensing data may cause a system malfunction or even a blackout. Big data analytics can be applied to quickly find anomalies from sensing data. A supervisory control and data acquisition (SCADA) system is a centralized monitoring and control capability for the power grid. SCADA still plays a role in smart grid; nonetheless, it will be upgraded to distributed monitoring system to have better scalability. Other more advanced communication infrastructures and monitoring systems (e.g. the AMI and the WAMS) are being deployed to monitor and control the grid.

System monitoring may include bus voltages, bus real and reactive power injections, electrical waves, etc. These measurements are stored in a controlling system for analysis. State estimation is used in system monitoring to estimate the power grid state through analysis of the measurements. The output of state estimation reveals potential operational problems in the smart gird in real time (or near real time). Actions can be taken to avoid problems (especially cascading failures) and potential side effects of those issues. Sensing data that reveals the operational status of the power grid is generated in real time (e.g. the PMU generates data at a high frequency, e.g. 60 to 120 frames per second for 60 Hz systems) [50]. It is impractical to apply complex network security schemes to protect sensing data. Therefore, state estimation using machine learning, data mining, and other techniques that are based on observed data can be a better solution.

8.3.2 Current Problems and Techniques

State estimation in the smart grid uses power flow models [141, 142]. A power flow model is a set of equations that depict the energy flow on each transmission line of a power grid. Using the DC power flow model, one of the widely accepted models is represented by a linear regression model as follows:

$$\mathbf{z} = \mathbb{H}\mathbf{x} + \mathbf{e}, \tag{8.8}$$

where $\mathbf{x} \in \mathbb{R}^{D}$ is set of state variables, $\mathbf{z} \in \mathbb{R}^{N}$ is the set of of measurement variables, $\mathbb{H} \in \mathbb{R}^{N \times D}$ is the measurement Jacobian matrix, and $\mathbf{e} \in \mathbb{R}^{N}$ is the measurement errors, which is assumed to have independent components.

Three statistical estimation criteria are commonly used in state estimation.

1) The maximum likelihood (MLE) criterion. Assume as given a sample $\mathbf{x}$ of observations that come from a distribution with an unknown probability density function. The function belongs to a certain family of distributions $\{f(\cdot|\theta)\}$, where $\theta\{\theta_1, \theta_2, \ldots\}$ is an unknown vector. The joint density function of $\mathbf{x}$ given θ is $f(\mathbf{x}|\theta)$. If the observed value $\mathbf{x}$ is considered to be fixed and θ is the variable, then the function is redefined as a likelihood function:

$$\mathcal{L}(\theta; \mathbf{x}) = f(\mathbf{x}|\theta). \tag{8.9}$$

The MLE is to find $\hat{\theta}$ that maximizes the likelihood function, as shown in Eq. (8.10).

$$\hat{\theta} = \arg\max_{\theta} f(\mathbf{x}|\theta) \tag{8.10}$$

2) The weighted least-square (WLS) criterion. Assume as given a set of points (x_i, y_i), $i = 1, 2, \ldots$, where x_i is an independent variable and y_i is a dependent variable whose value is found by observation. A function $f(x_i; \beta)$ is modeled to predict y_i. In most cases, there is a difference between the predicted value and the observed value. The difference is defined as a residual, so that $r_i = y_i - f(x_i; \beta)$. The WLS is to find $\hat{\beta}$ that minimizes the weighted square estimates.

$$\hat{\beta} = \arg\min_{\beta} \sum_{i=1}^{n} w_i[y_i - f(x_i; \hat{\beta})]^2, \tag{8.11}$$

where w_i is the weight of the i-th point.

3) The minimum variance (MV) criterion. Assume as given a function $f(\cdot)$ that predicts a dependent variable based on observation $\mathbf{x}$. The MV is to minimize the variance

$$\hat{\theta} = \arg\min_{\theta} \sum_{i=1}^{n} \text{var}\,(f(x|\theta)). \tag{8.12}$$

In most of the research work, the error $\mathbf{e}$ is assumed to be normally distributed with zero mean; all the three criteria mentioned above lead to an identical estimator (i.e. minimum mean square error (MMSE) estimator) with the matrix as follows [141, 142]:

$$\hat{\mathbf{x}} = (\mathbb{H}^T \mathbf{W} \mathbb{H})^{-1} \mathbb{H}^T \mathbf{W} \mathbf{z}, \tag{8.13}$$

where $\mathbf{W}$ is a diagonal matrix whose elements are reciprocals of the variances of errors, that is, $\mathbf{W} = diag(\sigma_1^{-2}, \sigma_2^{-2}, \ldots, \sigma_m^{-2})$, where σ_i^2 is the variance of the i-th measurement error.

If an attacker injects a false data vector $\mathbf{a} \in \mathbb{R}^N$ (nonzero) into the measurements, the resulting observation model is changed as follows:

$$\hat{\mathbf{z}} = \mathbb{H}\mathbf{x} + \mathbf{a} + \mathbf{e}. \tag{8.14}$$

To detect an attack, the measurement residual is examined in l_2-norm $\rho = \|\hat{\mathbf{z}} - \mathbb{H}\hat{\mathbf{x}}\|_2^2$. Given an arbitrary threshold $\tau \in \mathbb{R}$, if $\rho > \tau$, then a measurement is declared attacked. Several detection models using statistical learning methods have been studied, including supervised and semisupervised learning methods. Nonetheless, online learning methods can be more important for practical applications because the measurements are observed in real time in the smart grid. A general online learning system uses only the given samples at each algorithm's processing time. Energy fraud detection may not need to occur online or in real-time. Decision trees and artificial neural networks are commonly used as learning algorithms for energy fraud detection [140, 143].

8.4 Cloud Computing in the Smart Grid

As discussed before, the smart grid is halfway to the big data era. However, big data anaytics requires enormous computing and storage resources to extract useful information efficiently. How can utility companies achieve price/energy forecasts, attack detection, and other controls with a reasonable budget? A feasible solution is to introduce cloud computing into the ICT infrastructure. Cloud computing is a readily available resource and relatively cheap to use compared with dedicated computing centers.

8.4.1 Basics of Cloud Computing

Cloud computing is a model for enabling ubiquitous, convenient, on-demand network access to a shared pool of configurable computing resources that can be rapidly provisioned and released with minimal management effort or service provider interaction [144]. The cloud computing environment provides various services models such as software as a service (SaaS), platform as a service (PaaS), infrastructure as a service (IaaS), and network as a service (NaaS).

- Cloud software as a service (SaaS). In SaaS, the service provider manages and controls the underlying cloud infrastructure, including networks, servers, operating systems, storage, etc. Even the applications are provided by the service provider. A customer simply accesses the applications from client devices through a client interface (e.g. a web-based client).
- Cloud platform as a service (PaaS). Similar to SaaS, the service provider manages and controls the underlying cloud infrastructure in PaaS. However, a customer is provided the capability to create applications using programming languages and tools that are supported by the service provider.
- Cloud infrastructure as a service (IaaS). In IaaS, the service provider manages and controls the underlying cloud physical infrastructure. A customer is provided the capability to manage and control storage, networks, and some other fundamental computing resources. A customer is also able to run arbitrary software, including operating systems and applications.
- network as a service (NaaS). In NaaS, the service provider delivers virtual network services over the Internet. A customer uses NaaS based on a pay-per-use or monthly subscription basis.

There are different kinds of cloud deployments, including private cloud, public cloud, community cloud, and hybrid cloud. A private cloud in the smart grid is deployed and maintained by utility companies. A public cloud is provided by a service provider, and utility companies pay for the service when needed. A hybrid cloud is a combination of private clouds and public clouds.

Although security can be insured by a private cloud communicating with private networks, the deployment and maintenance costs are high. Using a public cloud alone in the smart grid may not be acceptable due to security concerns. A hybrid could balance security/performance and cost. Thus, the smart grid should deploy a hybrid cloud for big data analytics and other necessary applications.

8.4.2 Advantages of Cloud Computing in the Smart Grid

As sensors become integrated into virtually every piece of equipment in smart grid communications infrastructure, it would not be feasible or cost-effective to provide centralized data processing at utilities. Cloud computing brings three major advantages to the smart grid.

- First, there is no need to invest in the whole infrastructure, because only the infrastructure for the private portion of the cloud is deployed by utility companies. The resources from the public portion of the cloud can be rented at a relatively low price. In many cases, cloud computing uses a pay-as-you-go pricing model. The maintenance cost is also low, because it applies only to the private portion.
- Second, it is easier to implement applications in the cloud. Public cloud computing has virtually unlimited resources. Therefore, the utility companies need not worry about upgrading capacity for a large-scale system, which is no doubt a huge concern in the smart grid. The infrastructure in the cloud can also be rescaled according to adaptive requirements. Moreover, because of the cloud's elasticity, feature updates/upgrades can be done in a short amount of time without disturbing users or requiring them to install major updates or extra packages.
- Third, it is easier to access cloud service from a variety of smart devices. Because of that, monitoring and controlling the grid can be more flexible. Once security is provided for the transmission, a cloud service can be accessed from virtually anywhere with appropriate authorization.

Even with cloud computing, only significant data or processed data needs be communicated to the centralized data processor. In other cases, utilities may establish local data centers to serve as their private clouds. Even further, direct connections among edge nodes can be established and form fog computing to preprocess some data.

8.4.3 A Cloud Computing Architecture for the Smart Grid

In this subsection, we propose a a high-level cloud computing architecture for the smart grid, as shown in Figure 8.5. For simplicity, the cloud is a whole picture of both the private and public portions of the hybrid cloud. The input data of the cloud includes metering data and other sensing data from the grid, as well as weather forecasts, information from social networks, and other useful data from outside. Different types of cloud services (e.g. SaaS) are provided in the cloud for different applications. The components of the cloud include data management systems, data storage systems, and processing tools.

Much of the data management (e.g. data cleansing) is performed in the private portion of the cloud service to protect privacy of customers.

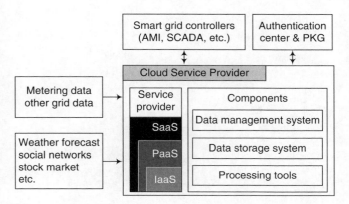

Figure 8.5 A cloud computing architecture for the smart grid.

The data storage system is separated in both the private and the public portions of the cloud, while the public part provides the most storage resources for big data. The processing tools for big data analytics, machine learning, and other techniques are mainly deployed in the public portion of the cloud to take advantage of the computing power available in the public cloud service.

8.5 Summary

In this chapter, we showed that big data is inevitable in smart grid communications. Applications such as demand response and modern control need the information provided by massive amounts of data. In particular, price and energy forecasts will be produced by big data analytics. Preliminary results were given to demonstrate the importance of big data analytics in energy forecasts. Besides demand response, attack detection in smart grid communications (especially monitoring networks) is critical for secure and efficient grid operations. Machine learning and data mining, as well as big data analytics, are useful tools to achieve attack detections for both metering data and sensing data. Thus, energy fraud and system failures can be prevented. Cloud computing was introduced in the chapter also as a way to accommodate big data analytics. An architecture with a hybrid cloud was proposed for that purpose.

9

A Secure Data Learning Scheme for Big Data Applications in the Smart Grid

In this chapter, a secure data learning scheme is proposed for big data applications in the information and communication technology infrastructure of the smart grid. The proposed scheme allows multiple parties to find the predictive models from their overall data, while not revealing their own private data to one another at the same time. Instead of deploying a centralized data learning process, the scheme distributes data learning tasks to the local learning parties as their own local data learning tasks to learn the value from local data, thus preserving privacy. In addition, an associated secure scheme is proposed to guarantee the privacy of learning results during the information reassembly and value-response process. An evaluation is performed to verify the privacy of the training data set, as well as the accuracy of learning weights. A case study is presented based on an open metering data analysis.

9.1   Background and Related Work

9.1.1   Motivation and Background

Nowadays, information and communication technology (ICT) infrastructure has been able by modern control techniques to tremendously improve the efficiency, reliability, and security of information systems [145]. Big data is an emerging topic due to its various applications [146]. However, it would not be so prevalent without the underlying support of the ICT, due to its extremely large volume of data and computing complexity. When a huge volume of data is

Smart Grid Communication Infrastructures: Big Data, Cloud Computing, and Security,
First Edition. Feng Ye, Yi Qian, and Rose Qingyang Hu.
© 2018 John Wiley & Sons Ltd. Published 2018 by John Wiley & Sons Ltd.

quickly generated, it is crucial to process, aggregate, store, and analyze such a massive amount of data in real time [147, 148].

For large data sets acquired by ICT companies, the overall data set does not always reside on the same database; instead, it may reside on disparate databases among various locations of companies. The challenge is that data analytics needs to deal with the big data problem. Many ICT companies do not have the infrastructure to support such needs. In terms of possible approaches, a data-learning task would be carried out by a decision maker from an ICT company. We categorize it into a centralized learning task and a distributed local learning task. Currently, most schemes are designed and dedicated to manage traditional small-scale amounts of data in a centralized approach. Figure 9.1 shows an example of a centralized learning process for big data applications. However, due to the performance bottleneck of a central hub, access to the data on the central hub is relatively inefficient in the traditional scheme. Data preprocessing could be relatively slow, and information privacy could be easily jeopardized. The central hub is not being released to enable performing heavily loaded analytics on massive amounts data in the traditional approach. Meanwhile, few of the schemes have effectively incorporated strong security and privacy protection measures during the process of big data learning among multiple entities. Security and privacy are without doubt the top concerns, no matter which learning approach a decision maker adopts, because of the crucial information about users' personal data and usage information the aggregated data may contain.

To close the gap, this chapter proposed a secure data learning scheme chapter for multiple entities in an environment of big data applications. It focuses on improving the disadvantages of traditional centralized learning tasks with large data sets. Specifically, the scheme

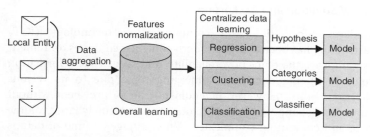

Figure 9.1 Traditional centralized learning process for big data applications.

can push the major learning processes from the site of a central hub to the sites of multiple learning entities (i.e. every origin of aggregated data) in a way that reduces the communication overhead. Moreover, based on the current learning algorithms, a penalty term is enforced on the objective function to avoid the overfitting problem, which a learning scheme may encounter while solving common regression problems. Additionally, we design an associated secure scheme with privacy preservation and identity protection to obtain the final learning result. The major security objectives of our proposed scheme are as follows:

1) Privacy. During the data learning process, only learning results are revealed by learning entities; learning results need to be controlled to maintain robust security against statistical attacks.
2) Malicious learning entity detection. The secure learning scheme should be able to detect when an entity has been compromised and subject to a malicious attack, so the system can maintain a high level of reliability.

9.1.2 Related Work

Several learning algorithms have been proposed to target the privacy and security issues in the data learning process. For example, in [149], the authors studied a privacy-preserving secure 2-entity computation framework based on linear regression and classification. In [150], based on constructing artificial neural networks, the authors developed a deep learning system to enable multiple entities to learn an objective without sharing their own data sets. In [151], the authors surveyed the basic paradigms of secure multiple entities computation and discussed their relevance to the field of privacy-preserving data mining schemes. In [152], the authors discussed approaches for privacy-preserving learning of distributed data in different applications and presented a solution of combined components for specific privacy-preserving data learning applications. In [153], the authors presented a cryptographically secure protocol for privacy-preserving decision trees. In [154], based on a differential privacy model, the authors constructed a privacy-preserving naïve Bayes classifier in a centralized scenario, where data miners could access a data set in a centralized way, and the data miner could deploy a classifier on the premise that the private information of data owners could not be inferred from the classification model.

9.2 Preliminaries

In this section, the fundamentals of generalized linear learning models are illustrated. The security model for big data applications is also given in this section.

9.2.1 Classic Centralized Learning Scheme

In this study, we adopt a supervised learning algorithm to evaluate the centralized data learning process. Given a large data set $\mathcal{X}$, the objective of a centralized learning scheme is to minimize a cost function J, avoid the phenomenon of overfitting by penalizing heavily weighted parameters of complex models, and provide high accuracy of predicted $\hat{y}$. In real world applications, a preprocessed data set possessed by a learning entity is a matrix $\mathcal{X}_k$ with M observations (rows) and F features (columns).

Based on the applications, the main learning tasks for our proposed scheme are carried out under supervised learning algorithms. From the existing data set $\{(x_i, y_i) \in \mathcal{X} \times \mathcal{Y}\}$ as the training data set, our objectives are as follows:

- To find a proper hypothesis function $h_w \in \mathcal{H}$ to solve a regression problem, so that $\{h_w : \mathcal{X} \to \mathcal{Y}\}$.
- To minimize the cost function J (i.e. $\min J$) of this hypothesis function h_w such that $\{y_i \approx h_w(x_i)\}$.

Note that a hypothesis function h_w of a major supervised learning algorithm is mainly parameterized by the hypothesis weights $w \in \Theta$. In a normal case, a cost function $J(h_w(x), y)$ is applied to approximate how well a hypothesis function $h_w(x)$ performs when the true output is y. For a data set possessed by one learning entity, we define the "true" risk $\mathcal{R}(h_w)$ as

$$\mathcal{R}(h_w) = \mathbb{E}_p[J(h_w(x), y)] = \int \int p(x, y) J(h_w(x), y) dx dy. \qquad (9.1)$$

where p is the true distribution over x and y. However, risk $\mathcal{R}(h_w)$ cannot be computed directly, since the distribution p is unknown to the learning algorithm. Therefore, we apply empirical risk minimization [155] to compute an approximation of risk $\mathcal{R}$ by averaging the loss function on a single training data set:

$$\mathcal{R}_{\text{emp}}(h_w) = \frac{1}{M} \sum_{i=1}^{M} J(h_w(x), y). \qquad (9.2)$$

Finally, the learning algorithm selects a hypothesis function h_w^* that minimizes the empirical risk as the optimal learning model to perform centralized learning scheme. The hypothesis is defined as follows:

$$h_w^* = \arg \min_{h_w \in \mathcal{H}} \mathcal{R}_w(h_w). \tag{9.3}$$

9.2.2 Supervised Learning Models

9.2.2.1 Supervised Regression Learning Model

The quadratic cost function J for supervised linear regression algorithms is defined as follows:

$$\min J(Y, h(X)) = \frac{1}{2} \|Y - h(X)\|^2. \tag{9.4}$$

For instance, if linear regression is used as the training model to solve this problem, then the hypothesis function $h(x)$ for the training data set is presented as follows:

$$h(x) = \sum_{j=0}^{M} w_j x_j = w^T x, \tag{9.5}$$

where w_j is defined as the space of a linear function mapping from $\mathcal{X}$ to $\mathcal{Y}$ and x_j is the value of the j-th feature in a training sample. The quadratic cost function for the training data set is defined as follows:

$$R_{emp}(h_w) = \hat{\mathbb{E}} J(h_w(x) - y)^2. \tag{9.6}$$

The cost function is to measure the distance between a $h_w(x)$ to the corresponding y given a value w.

9.2.2.2 Regularization Term

For generalized linear learning models, the cost function J is usually treated as being convex [156]; thus a summation of multiple cost functions remains convex. In order to solve these specific optimization problems, we first transform them to unconstrained minimization problems. Readers should refer to [157] for detailed methods of solving an unconstrained problem.

With a relatively small size local data set, learning models may encounter an *overfitting problem* [158], especially when empirical risk minimization is adopted in the models. In order to prevent overfitting while minimizing the empirical risk $R_{emp}(h_w)$, a penalty term is needed based on the learning model to form an L_2 ridge regularization regression [159]. This process aims at minimizing the squares of the parameters. By performing an L_2 ridge regularization regression this penalty term would not only effectively prevent an overfitting problem but also regularize the unconstrained minimization problem. The penalty approach used in this scheme is defined as follows:

$$\min_{h \in \mathcal{H}} \frac{1}{C} \sum_{i=1}^{C} J(h_k(w,x),y) + \lambda P(h_k),$$

$$\Rightarrow \min_{h \in \mathcal{H}} \frac{1}{C} \sum_{i=1}^{C} (h_k(w,x) - y)^2 + \frac{\lambda}{2} \|w\|_2^2, \tag{9.7}$$

where λ is treated as the penalty coefficient of the L_2 ridge penalty term $\|w\|_2^2$. Normally an unconstrained minimization problem with a convex function can be solved by existing descent optimization algorithms, e.g. the gradient descent method and the classic Newton's method. However, the learning rate η used in a descent method is hard to choose. In this study, we are inspired by the stochastic gradient descent method [160] to perform fast computation.

9.2.3 Security Model

The security model used for analysis is a semimalicious model, which is a subcategory of secure multiple learning entities computation [161]. In a semimalicious model, all the entities are preregistered with the central hub. In particular, the central hub is a trusted learning entity, while some of the entities are treated as potential malicious identities that are compromised by any adversarial group. A compromised and malicious learning entity may perform as follows:

1) Refuse to participate in the learning scheme.
2) Deliberately substitute or falsify its local learning result.
3) In the worst case, viciously abort the learning scheme prematurely.

From the discussions mentioned before, there are two major security and privacy objectives in designing such a learning scheme: (i) During the learning process, only the final training result is revealed with protection; local data information should remain private under any circumstances. And (ii) during the learning process, a compromised and malicious learning entity would be detected by other normal entities as the learning scheme is being adopted. In the next section, we will illustrate how the secure learning scheme will perform the distributed data learning tasks while identifying malicious entities and guaranteeing the security and privacy of data information from being viciously exposed to others.

9.3 Secure Data Learning Scheme

In this section, we present the secure data learning scheme. We first introduce the proposed data learning scheme, with a focus on the algorithm design. We then introduce the associated security scheme with respect to data privacy and identity protection.

9.3.1 Data Learning Scheme

A learning entity P_k fits its hypothesis function h_k when a data set of $\{\mathcal{X}_k = (x_1, x_2, \cdots, x_M)^T \in \mathbb{R}^{M \times F}\}$ is arrived during a fixed time interval T. For this learning entity P_k, its cost function J_k of the hypothesis function h_k could be denoted as $J_k(h_k(\mathcal{X}_k))$. In light of the principles of descent algorithms, during the τ-th iteration of this learning study, we formulate this as a minimization problem $f(x)$ for a learning entity P_k, with the objective of minimizing the cost function,

$$\min_{h \in \mathcal{H}} f(x) = J^\tau(Y, h^\tau(X)) + \frac{\lambda}{2} \|w\|_2^2. \tag{9.8}$$

In the beginning, each learning entity only trains its own private local data to fit its own training model. The local training model is a relatively suitable model for the local learning entity; however it would most likely be a poor fit as a training model for the overall data set. Hence, the objective becomes to solve J and to find the optimal value of the hypothesis weights vector w^* from h_k and λ_k^*.

In order to find the optimal value of the hypothesis weights vector w^* from h_k and λ_k^*, we need to find the solution to Eq. 9.8 by calculating the partial derivatives of w. Note that some of the learning models we discussed before can be applied to the hypothesis function to solve certain regression problems. As a general process, during the τ-th data learning iteration, the learning model of entity P_k will update its hypothesis function $h^\tau(\mathcal{X}_k)$ based on the empirical risk minimization discussed before. Inspired by the stochastic gradient descent algorithm, a detailed learning algorithm to obtain hypothesis weights vector w_k^τ for an entity P_k in τ-th iteration is presented in

---

Algorithm 9.1 Learning algorithm for obtaining hypothesis weights vector w_k^τ

1: Initialize the weights w;
2: **for** $k = 1 : N$ **do**
3:    **for** $i = 1 : M$ **do**
4:       Set the learning rate η accordingly
5:       Update the weights $w' := w - \eta \frac{\partial J}{\partial w}$
6:       Obtain the current optimal vector w'
7:       Compute $w = w' + \frac{\lambda}{2}\|w\|_2^2$
8:    **end for**
9:    Update w with $w_k := w_k - \eta \frac{\partial J}{\partial w} - \eta \cdot \lambda_k \cdot (w_k)$
10:    Jump to the next iteration until it stops
11: **end for**
12: **return** the final hypothesis weight vector w^*

Algorithm 9.1. For simplicity, we use w as the learning weight vector for a local training data set $\{\mathcal{X}_k = (x_1, x_2, \cdots, x_M)^T \in \mathbb{R}^{M \times F}\}$, where $i \in \{1, \ldots, M\}$, $j \in \{1, \ldots, F\}$, J and h for the cost and hypothesis function of P_k during the τ-th iteration, respectively.

In Algorithm 9.1, w' represents the current optimal weight before performing the L_2 ridge penalty term. For different supervised learning models, there are different hypothesis functions for h. For example, if a polynomial regression problem is encountered, the gradient of cost function J is defined as $(y_i - h(x_i))x_{ij}^p$, where in this case (i,j) represents the index of each observation and feature of a data set. If a classification problem is encountered, then the gradient of a 0–1 logistic regression cost function J can be defined as $(y_i - h(x_i))x_{ij}$. When we take a look into the hypothesis weights vector $w := w - \eta \frac{\partial J}{\partial w}$, we can see that the output of $\frac{\partial J}{\partial w}$ depends on the supervised learning algorithm. It minimizes the cost function and makes the training result better fit the learning model.

9.3.2 The Proposed Security Scheme

9.3.2.1 Privacy Scheme

In this part we present the proposed privacy scheme. As shown in Figure 9.1, the step of feature normalization towards a large aggregated data set is crucial. When features differ by orders of magnitude, a standard feature normalization process performs feature scaling so that descent algorithms can converge faster. During the aggregation process, our objective is to find an associated privacy scheme to incorporate with the feature normalization process.

For a joint data set, the goal of Z-score normalization is to rescale the features so that the properties of a standard normal distribution with $\mu = 0$ and $\sigma = 1$ can be found and the standard scores of the joint data set could be calculated $z = \frac{x-\mu}{\sigma}$ for each data entry. By subtracting the mean value of each feature from the data set and scaling the feature values by their respective standard deviations, the exact value in each data entry is in a range of $[-1, 1]$. In light of this, the exact values of data content from each observation and feature have been normalized, and they are different from their original data. By performing Z-score normalization, we could expedite the process of data learning when the sample size and the number of learning entities increase. After the Z-score normalization, the local value V_k can also be a $M \times F$ based matrix.

Z-score normalization has modified the original data content from the malicious users. However, in order to fully achieve the goal of preserving data privacy, it is necessary to prevent exposure of the real size of the data set. A noise *ns* between 0 and 1 is added to increase security robustness against statistical analysis during normalization in the learning entity P_k. When the features are normalized, it is quite important to store the related values we used for normalization computation. Otherwise, after obtaining the parameters from the model, malicious attackers would still be able to recover the raw values, perform predictions, and compromise the integrity of the data. A set-up noise would be a good disguise against those passive attacks.

9.3.2.2 Identity Protection

If a learning entity P_k is compromised by a malicious entity, the trusted centralized hub should be able to detect the abnormality immediately. Once a P_k is compromised, it normally will not reveal its malicious identity to the central hub but will keep its secret malicious identity anonymous and deliberately falsify or substitute its local learning result to the central hub for the purpose of jeopardizing the learning scheme and the whole overall learning result. Therefore, our objective is to determine whether the identity of a P_k is legitimate or not, without requesting any identity information from a malicious P_k itself, and to report the malicious P_k to the system administrator.

Specifically, a fast security solution scheme inspired by the zero-knowledge proof [162] is proposed. A P_k first computes the 512-bit hash value of its local learning result $\widehat{y_k}$ during τ-th iteration. P_k then combines it with a time-stamp tsp as a joint message $M_1 = \mathbf{H}(\widehat{y_k}) \| tsp$, encrypts the ticket using its own private key P_k^-, and then sends the encrypted message $M_2 = \mathbf{E}(M_1; P_k^-)$ to the central hub. Once the central hub receives the encrypted ticket, it sends a 512-bit hash value of the current overall learning value $M_3 = \mathbf{H}(\mathbf{V})$ back to P_k. After receiving it, P_k concatenates the joint message of M_1, M_3, and the updated time-stamp tsp' together as $M_4 = M_1 \| M_3 \| tsp'$, encrypts it with its private key P_k^-, and sends it as $M_5 = \mathbf{E}(M_4; P_k^-)$ to the central hub again. Since all the entities are preregistered, the central hub then looks up P_k's public key P_k^+ from a public keys repository for all entities located at the trusted side. Once P_k^+ is obtained, the central hub could decrypt the encrypted joint message M_5 to perform the identity check and determine whether P_k is truly legitimate or not. If P_k is legitimate, the central hub allows the access of P_k so that P_k submits its local learning result to the central hub during τ-th iteration.

The proposed security scheme based on zero-knowledge proof is illustrated in Figure 9.2. Note that to address security concerns, the scheme should be processed at the beginning of each data learning iteration, so that any abnormal operation occurring in an iteration can be detected. A nice property of this zero-knowledge proof based

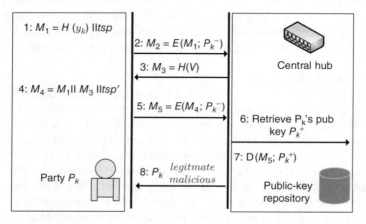

Figure 9.2 Proposed security scheme based on zero-knowledge proof.

scheme is that one entity could be easily verified to possess certain information simply based on revealing its basic public information. Through the illustration, one entity's challenge process could be proven without revealing secret information. A security analysis of this associated security scheme is presented in 9.3.4 to demonstrate its security robustness.

9.3.3 Analysis of the Learning Process

In the proposed algorithm, we repeatedly run through the data set, and each time we encounter a learning example, we update the parameters according to the gradient of the error term $(y - h_w(x))$ with respect to that single learning example only. It could immediately start to perform the learning process and keep track of it with each observation. As we discussed in section IV.A, inspired by the stochastic gradient descent, it could reach the optimal weight much faster and is a reasonably good choice when facing a large data set.

We would also expect to find out that as the number of iterations increases, the minimized value cost function will converge. This also means that for a convex cost function J, the hypothesis function would be $h^\tau(X) = h^{\tau-1}(X)$, and $h^\tau(X) = Y$ with a very small of tolerance.

9.3.4 Analysis of the Security

In order to fully achieve the goal of preserving data privacy, the protection of noise ns is crucial in terms of generation and computation during the Z-score normalization. Here we consider ns generated from an irreversible hash function with the input $\|ID\|\|tsp\|^2$ and convert it into a decimal value $d = H_D(\|ID\|\|tsp\|^2)$. Then the noise ns would be fixed into the range of 0 and 1 by performing $ns := \frac{d}{|d|}$.

During the verification process, a reverse computation could be adopted to check on the L_2 form of weights w and the decimal hash value. Since d is computationally irreversible, the range of noise ns could be guaranteed in this case. When it comes to the anonymization of data set sample size, ns would play an important role in incorporating the anonymity of a local value. Even if an adversary obtains the dimension information of a possible local value, it would still be in vain, since ns is not computationally accessible to the adversary. Additionally, ns being in the range of 0 and 1 still holds a good property of computational efficiency, as inherited from the Z-score normalization.

9.4 Smart Metering Data Set Analysis—A Case Study

In this section, we describe a case study on using the proposed secure data learning scheme to solve the regression problem of smart metering data sets from the UMass Trace Repository [163]. We first introduce the networking model of a smart grid AMI, then perform regression work on the data sets.

9.4.1 Smart Grid AMI and Metering Data Set

As one of the most critical infrastructures, the smart grid has adopted advanced and powerful ICT to improve all sorts of aspects of the power grid [7, 164–171, 147]. In a smart grid advanced metering infrastructure (AMI), as shown in Figure 9.3, a data aggregation point (DAP) aggregates massive amounts of smart metering data from distributed smart meters at each household, preprocesses the aggregated data, and forwards the preprocessed data to the operation center via the smart grid's wide-area networks (WAN) through a local master gateway of the designated AMI concentrator. This high volume of collected data, which contains the summary of power usage and personal behavior patterns, is analyzed by a big data platform and metering data management system (MDMS) for accurate real-time control and optimized resource allocation.

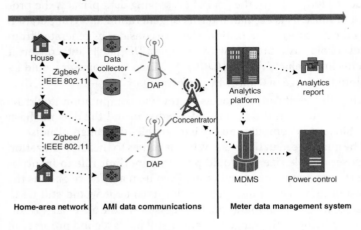

Figure 9.3 The ICT architecture of AMI in the smart grid.

Table 9.1 A sample set of smart metering data.

Name	Meter #	Current	Previous	⋯	Total (KWh)
Suite 101	45600	321.2	485.2	⋯	258.7
Suite 102	45601	320.8	483.9	⋯	341.0
Suite 103	45602	322.7	486.1	⋯	493.2
⋯	⋯	⋯	⋯	⋯	⋯
Suite 899	47212	320.3	482.9	⋯	234.6

An example smart metering data set, shown in Table 9.1, is based on a frequent metering report from a smart meter. Normally in order to make efficient energy-buying decisions based on usage patterns, to perform power theft detection, and to correct its service performance, utility companies would have a corresponding matrix $\mathcal{Y}$ with the same M observations and one feature, for example, power usage quality. As stated in the system model, a DAP D_k trains its hypothesis function $h_k(w, x)$ when a data set of $\{\mathcal{X}_k = (x_1, x_2, \cdots, x_{M \times s_k})^T \in \mathbb{R}^{(M \times s_k) \times F}\}$ arrives during a a fixed time interval T_a. During this T_a, each smart meter reports the metering data to its own DAP. Since each DAP governs a different number of smart meters, we define the DAP D_k that governs s_k number of smart meters. Therefore, the total data matrix that a DAP D_k receives during a time interval T_a has $M \times s_k$ observations and F features. Considering the fact that all the metering data is reported in a fixed short time interval, the sample size is likely to be relatively small. As discussed in "9.2.2.2", specifically the second paragraph of "9.2.2.2", this might lead to an overfitting problem for learning models, especially when empirical risk minimization is adopted.

In a DAP D_k, the cost function could be considered as $J_k(h_k(w, x))$. In the study case of the τ-th iteration, we formulate this as a minimization problem for a DAP learning entity D_k, with the objective function

$$\min_{h_w \in \mathcal{H}} \frac{1}{C} \sum_{i=1}^{C} J_k(h_k(x), y) + \frac{\lambda}{2} \|w\|_2^2, \tag{9.9}$$

where C represents the total observations $C = M \times s_k$ of a joint data set among N DAP learning entities during time interval T_a, and i represents the i-th observation in the data set. In this study, the concentrator is surrounded by N DAPs forming a mesh-based graph model [166], for the requirement of robust and reliable neighborhood area communications [167].

9.4.2 Regression Study

A regression study is conducted based on the data sets from [163]. In this scenario, 30 data learning entities are involved in training their own local data, and a central hub-based model is established to perform the proposed scheme. Based on the data set, metering information is reported every five minutes during a day. Several features, such as temperature, humidity, wind degree and direction, etc., are contained in the environmental data set forming the X, while the power usage is recorded as Y with the unit of watt. Six features, including "insideTemp," "outsideTemp," "outsideHumidity," "wind-Speed," "windDirectionDegrees," and "windGust," are selected as the learning features, and a ridge-based regression task is conducted among all the 30 local entities.

Without loss of generality, we show the plots of hypothesis weights of learning entities numbers 1, 5, and 6 in Figure 9.4, Figure 9.5, and Figure 9.6 respectively, with respect to the ridge coefficients λ. In each figure, six lines represent the chosen six features of the coefficient vectors. In the beginning, each weight vector w from different learning entities is randomly sampled from a Gaussian distribution, as presented in each figure with different starting points. While λ tends toward a certain threshold, weights found by the regression models from different learning parties start to become stabilized towards the sampled vector w. Meanwhile, the regularized weights are getting lower and will eventually converge to zero.

Once the initial hypothesis weights from the ridge-based regression are obtained locally, the iteration process starts as well. By adopting the modified gradient descent algorithm, the values of the cost function J_i decrease as the iterations move on in Figure 9.7. After approximately 50 iterations, we achieve convergence among all the learning entities. As a comparison, the overall data set aggregated through the aggregation protocol and processed through centralized data learning scheme remains a large cost, because sample size of the overall data set is larger than the local data set.

Moreover, we can see that the cost value in each function J_i is quite large. The results can be improved if the nonlinear regression model is carefully chosen. On the local side, each entity performs with almost

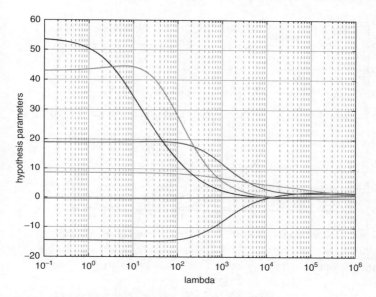

Figure 9.4 Regularization results using learning entity 1.

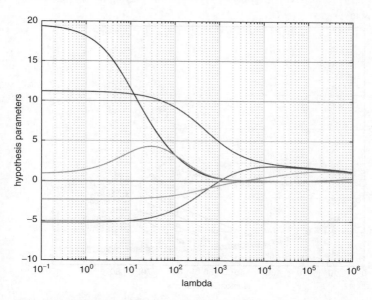

Figure 9.5 Regularization results using learning entity 5.

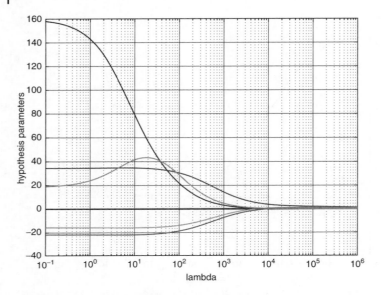

Figure 9.6 Regularization results using learning entity 6.

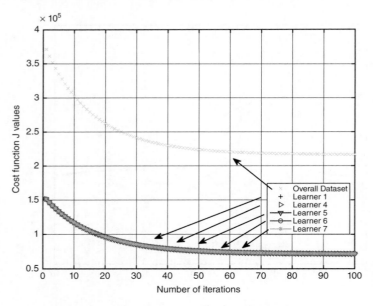

Figure 9.7 Convergence on the values of cost functions J_i.

the same cost, and it actually explains the frequency-based power consumption follows a regular amount of data size. In summary, it follows the nature of descent algorithms: when the value of cost function J_i converges, the hypothesis weights w_i converge.

9.5 Conclusion and Future Work

In this chapter, we presented a secure data learning scheme for multiple entities in an environment of big data applications. We considered the scenario where multiple entities, which act as data holders, were intended to find predictive models from their overall data. We proposed a scheme in which the property of scalability of data learning is used as an alternative computation technique to provide security to multiple entities. The proposed scheme was inspired by descent algorithms and considered ways to avoid the overfitting problem. Additionally, an associated secure scheme was also proposed to secure the privacy of local information against content leakage and statistical attacks, as well as the protection of identities. Both theoretical learning and security analysis are provided by this scheme. A case study was conducted by adopting the proposed scheme to perform a regression problem and investigate the smart metering data in the smart grid. The proposed scheme was targeted at improving the potential computation and security drawbacks of the centralized training task of big data applications. Future research in this area may consider massive data sets and other supervised learning models to provide more accurate predictions.

10

Security Challenges in the Smart Grid Communication Infrastructure

In this chapter, we will discuss security challenges in the smart grid communication infrastructure. The requirements for building a reliable smart grid communications network will be described, especially the requirements for the utility's private and public networks used in the smart grid.

10.1 General Security Challenges

The smart grid communication infrastructure is complex and evolving. Security challenges in the smart grid are continuously changing, depending on a particular system. In the transition from the traditional power grid to the smart grid, many legacy systems need to be protected before being upgraded. In the smart grid, there are a large number of end points in many geographic locations. In addition, most systems in the smart grid are always required to be online. Besides these technical challenges, other security challenges may come from the culture of security through obscurity and a lack of standards and regulations.

10.1.1 Technical Requirements

The smart grid communication infrastructure represents a technical challenge that is far beyond the simple addition of an information technology infrastructure on top of an electrical network. The number of widely distributed nodes that are tightly coupled and operating in the electrical network has grown over many years. It is very

Smart Grid Communication Infrastructures: Big Data, Cloud Computing, and Security,
First Edition. Feng Ye, Yi Qian, and Rose Qingyang Hu.
© 2018 John Wiley & Sons Ltd. Published 2018 by John Wiley & Sons Ltd.

challenging to figure out where intelligence needs to be added. Another challenge comes from the continuous operation of the current power grid. The smart grid implementation will be a continuous evolution of successive projects over many years. Incorporating a huge number of legacy systems will pose a constant challenge to the evolution of the smart grid. Besides, different stakeholders are responsible for different parts of the system. Independently, each may make different choices about the evolution and use of the grid.

The functional requirements of major applications in the smart grid are listed in Table 10.1. It is clear that the requirements vary from one application to another. There is no single solution to all the challenges in smart grid applications. For example, the AMI and the monitoring infrastructure have completely different requirements, from security to backup power. The diversity of requirements further increases the complexity of the technical challenges in the smart grid.

Table 10.1 Functional Requirements.

Application	Security	Bandwidth	Reliability	Latency	Back-up Power
AMI	High	14–100 Kbps	99.0–99.99%	2000 ms	0–4 hrs
Meter Data Management	High	56 Kbps	99.00%	2000 ms	0 hr
DR	High	56 kbps	99.00%	2000 ms	0 hr
DLC	High	14–100 Kbps	99.0–99.99%	2000 ms	0–4 hrs
Distributed Generation	High	9.6–56 Kbps	99.99%	2000 ms	0–1 hr
Charging PHEV	Medium	9.6–56 Kbps	99.90%	2000 ms–5 min	0 hr
Emergency Response	Medium	45–250 Kbps	99.99%	500 ms	72 hrs
Outage Management	High	56 Kbps	99.00 %	2000 ms	0 hr
Transformer Monitoring	Medium	56 Kbps	99.999%	500–2000 ms	0 hrs
Voltage Monitoring	Medium	56–10 Kbps	99.999%	2000–5000 ms	0 hrs

10.1.2 Information Security Domains

Security in smart grid communications infrastructure can be divided into different information domains as follows.

- Public supplier and maintainer domain
- Power plant domain
- Substation domain
- Telecommunication domain
- Real-time operation domain
- Corporate IT domain

Interdependencies among different information security domains present challenges when evaluating the impacts of a cybersecurity incident.

10.1.3 Standards and Interoperability

Utility providers are different even within the same country. The major challenge is to integrate interchangeable parts and technologies from a variety of providers worldwide. There is a need for interoperability standards to address this issue. Standards are also required to test the relatively new technologies that are applied to the smart grid communications infrastructure. One major challenge is the continuous operation of the power grid. The upgrading process to the smart grid will need to occur without interrupting critical grid operations.

10.2 Logical Security Architecture

In the guidelines for smart grid cybersecurity published by the National Institute of Standards and Technology (NIST) [172], a logical security architecture is proposed to describe *where*, at a high level, the smart grid needs to provide security.

10.2.1 Key Concepts and Assumptions

The logical security architecture specifies the following key concepts and assumptions:

- **Defense-in-depth strategy**. Smart grid cybersecurity should be applied in layers, with one or more security measures implemented at each layer. The objective is to mitigate the risk of one component of the defense being compromised or circumvented.

- **Defense-in-breath strategy.** Security activities are planned across the system, network, or subcomponent life cycle: product design and development, manufacturing, packaging, assembly, system integration, distribution, operations, maintenance, and retirement. The goal is to identify, manage, and reduce the risks of exploitable vulnerabilities across all parts of the life cycle.
- **Power system availability.** The primary focus of power systems engineering and operations is supporting the safe and reliable delivery of electricity. Existing power system designs and capabilities have been successful in providing this availability by protecting against inadvertent actions and natural disasters. These existing power system capabilities may be used to address the cybersecurity requirements.
- **Microgrids.** An implied hierarchy in availability and resilience eliminates potential peer-to-peer negotiations between microgrids. Microgrid models suggest that availability starts in a local microgrid and that resilience is gained by aggregating and interconnecting those microgrids. These interactions are not just theoretical. Microgrids are intended to operate either as islands or interconnected entities; islands are key where critical operations need to be maintained.
- **Wide-area situation awareness (WASA).** WASA is often shared between business entities; such information should be specified and secured in accordance with the principles of service-oriented architecture (SOA) security. Examples of such interactions might include the exchange of WASA between a provider and aftermarket consumers (co-op or aggregator), between a utility and emergency management, or between adjacent bulk providers.

A logical security architecture needs to provide protection for data at all interfaces within and among all smart grid domains. The logical security architecture baseline assumptions are as follows:

- A logical security architecture promotes an iterative process for revising the architecture to address new threats, vulnerabilities, and technologies.
- All smart grid systems will be targets.
- There is a need to balance the impact of a security breach and the resources required to implement mitigating security measures.

(Note: The assessment of the cost of implementing security is outside the scope of this chapter. However, this is a critical task for organizations as they develop their cybersecurity strategy, perform a risk assessment, select security requirements, and assess the effectiveness of those security requirements.)

- The logical security architecture should be viewed as a business enabler for the smart grid to achieve its operational mission (e.g. avoid rendering mission-purposed feature sets inoperative).
- The logical security architecture is not a one-size-fits-all prescription, but rather a framework of functionality that offers multiple implementation choices for diverse application security requirements within all electric sector organizations.
- As is common practice, the existing legacy systems will need to be considered as the new architecture is designed. Security implications will need to be reviewed and updated, both to consider the legacy security mechanisms and the current state of security practice.

10.2.2 Logical Interface Categories

A total of 22 logical interface categories are listed in the NIST guidelines for developing a cybersecurity strategy and implementing a risk assessment to select security requirements. This information may also be used by vendors and integrators as they design, develop, implement, and maintain security requirements. The logical interface categories are as follows:

1) Interface between control systems and equipment with high availability, and with compute and/or bandwidth constraints.
2) Interface between control systems and equipment without high availability, but with compute and/or bandwidth constraints.
3) Interface between control systems and equipment with high availability, without compute or bandwidth constraints.
4) Interface between control systems and equipment without high availability and without compute or bandwidth constraints.
5) Interface between control systems within the same organization.
6) Interface between control systems in different organizations.
7) Interface between back-office systems under common management authority.
8) Interface between back-office systems not under common management authority.

9) Interface with B2B connections between systems usually involving financial or market transactions.
10) Interface between control systems and noncontrol/corporate systems.
11) Interface between sensors and sensor networks for measuring environmental parameters, usually simple sensor devices. possibly with analog measurements.
12) Interface between sensor networks and control systems.
13) Interface between systems that use the AMI network.
14) Interface between systems that use the AMI network with high availability.
15) Interface between systems that use customer (residential, commercial, and industrial) site networks.
16) Interface between external systems and the customer site.
17) Interface between systems and mobile field crew laptops/ equipment.
18) Interface between metering equipment.
19) Interface between operations decision support systems.
20) Interface between engineering/maintenance systems and control equipment.
21) Interface between control systems and their vendors for standard maintenance and service.
22) Interface between security/network/system management consoles and all networks and systems.

10.3   Network Security Requirements

Readers may refer to the NIST guideline for detailed requirements of each logical interface category. In this section, we categorize all interfaces into two classes: utility-owned private networks and public networks in the smart grid. The security requirements are discussed based on the two classes.

10.3.1   Utility-Owned Private Networks

Data in smart grid communications is generated by many different intelligent devices together with direct input from human administrators for different purposes. The data transmitted over private networks can be categorized into four types, namely, metering data,

Table 10.2 Security requirements for data transmitted over private networks.

	Confidentiality	Integrity	Non-repudiation
Metering data	✓	✓	
Pricing/tariff information		✓	✓
Monitoring data		✓	
Control message		✓	

monitoring data, control messages, and pricing/tariff information. Strictly speaking, metering data is a kind of monitoring data, and pricing/tariff is part of control messages. However, metering data and pricing/tariff mostly contribute to demand response, while other monitoring data and control messages are mostly applied to other grid operations. The security requirements of those four types of data in private networks are summarized in Table 10.2.

Metering data is gathered from customers, in particular the power consumption of each household. Metering data contains much private information. For example, from the pattern of energy consumption, it is possible to sketch the lifestyle of a customer. Therefore, it is vital to provide confidentiality to metering data. In addition, integrity is also important to metering data. Manipulation of energy consumption (e.g. energy theft) may cause loss to the service provider. More importantly, manipulation of energy consumption data may cause the service provider to deviate from optimal control of the power grid, which in turn will lead to unnecessary fuel waste and pollution. However, non-repudiation may not be as critical as the other two security requirements for two reasons. 1) Providing non-repudiation, which usually is achieved by digital signature, may compromise the identity of the customer and thus jeopardize privacy. 2) Data in the uplink is frequently transmitted by simple devices such as smart meters or DAPs. They have limited computational capability, so applying public key cryptography frequently is not practical.

Pricing/tariff information is generated and transmitted from the service provider to customers in several ways. The most efficient way is through the private networks in AMI so that smart meters can receive real-time updates and adjust the power consumption of each smart appliance accordingly. For such transmissions, confidentiality can be dropped since pricing/tariff information is meant for all (or the majority) of the customers. Nonetheless, integrity and non-repudiation are critical requirements. Pricing/tariff information must remain fresh and correct all the time so that demand response can be applied accordingly. Customers (i.e. smart meters in this case) must be able to verify the legitimate sender (i.e. the service provider) so that forgery of such information can be detected, reported, and discarded. Besides, the availability of metering data is important but not critical, since alternative means for retrieving metering data can still be used. The types of security that could be applied are limited to the computational capabilities of a smart meter. Moreover, key management of millions of meters will pose significant challenges. Standard development is required to test the capabilities of new technologies used with smart meters.

The monitoring data of power grid status is gathered by low-profile sensors (e.g. PMUs). Obviously, data integrity needs to be provided so that the service provider can monitor the grid correctly. However, such sensors have limited computational power and power supplies. Moreover, monitoring data has strict latency requirements (e.g. about 10 ms for PMU data in WAMS). Therefore, it is not necessary to provide confidentiality and non-repudiation to monitoring data. However, integrity of monitoring data must be guaranteed for precise grid monitoring and optimal grid operations. Certain control messages to intelligent components (e.g. in response to hazardous situations) also require integrity. Due to low latency and limited computational power at the receiver side, confidentiality and non-repudiation may not be provided. Nonetheless, logs and files containing forensic evidence following events should probably remain confidential for both critical infrastructure and organizational reasons.

10.3.2 Public Networks in the Smart Grid

Different types of information are constantly transmitted over the public network in smart grid communications. General security requirements are listed in Table 10.3 for each type of information.

The security-related issues for the interface between external systems and the customer site (for example, between a third party and the HAN gateway) include confidentiality and integrity. Not all security services are required for this interface. Obviously, the pricing forecast does not need to remain confidential; nonetheless

Table 10.3 Security requirements for data transmitted over the public networks.

	Confidentiality	Data Integrity	Non-repudiation
Pricing forecast		✓	✓
Raw energy forecast	✓	✓	✓
Preprocessed data	✓	✓	✓
External information		✓	

its integrity and non-repudiation must be guaranteed. Preprocessed data is transmitted from local control centers to the cloud computing service. Big data analytics can be applied to such data to extract energy forecasts, and thus it is not meant for the public. Therefore, confidentiality, integrity, and non-repudiation are all required for preprocessed data. The raw energy forecast is made from big data analytics by the cloud computing service. Again, it is not meant for the public and thus confidentiality is required. Integrity and non-repudiation are also important for the raw energy forecast. External information is usually open to the public, so confidentiality is not required. Neither is non-repudiation, since the external sources may not even cooperate on this term. However, integrity should be provided. Availability and bandwidth are not generally critical between external parities and the customer site, since most interactions are not related to power system operations in real time.

10.4 Classification of Attacks

In this section, we will discuss *component-based attacks* and *protocol-based attacks* in the smart grid communications infrastructure.

10.4.1 Component-Based Attacks

Stuxnet was specifically programmed to attack SCADA in 2010 [173]. This malicious computer worm could reprogram programmable logic controllers, which allow the automation of electromechanical processes such as those used to control process plants and nuclear plants. The design and architecture of Stuxnet are not domain specific, and it could be tailored to become a platform for attacking modern SCADA systems of the power grid.

The **PMU** could suffer from three types of attacks [174]. A *reconnaissance attack* is defined as an attack that reconnoiters and identifies the system before an attack by a cyber-attacker. A *packet injection attack* is defined as sensor measurement injection and command injection. The third type of attack is denial of service. Since the PMU is required to have precise synchronization, another attack against the PMU is a *time synchronization attack* [175]. An example is the TSA-GPS spoofing attack, which is achieved by inserting a delay on satellite signals and not modifying them in the encoding process. The goal is to maximize alternations among the receiver's clock offset with and without the attack. The main functions of the PMU affected by TSA are fault detection in the transmission line and inaccurate event location.

The **SCADA** may suffer from internal and external attacks. Internal attacks against the SCADA may be launched by employees or contractors who have access to the system. External attacks are nonspecific malware and hackers. For example, Stuxnet could be launched as either an internal or external attack. Attacks launched by a former insider may target special knowledge of the SCADA system. Attacks launched by external hackers or terrorists may not target special knowledge. Natural or even man-made disasters should be considered attacks on the system.

Other cyberattacks can be launched against a specific component in the smart grid. For instance, regular cyberattacks against an SCADA system may include web server or SQL attacks, email attacks, zombie recruitment, DDoS attacks, etc. Some of the vulnerability points in the smart grid system could be unused telephone lines, use of removable media, infected Bluetooth-enabled devices, Wi-Fi-enabled devices that have an Ethernet connection to a SCADA system, insufficiently secure Wi-Fi, corporate web servers, email services, Internet gateways, etc.

10.4.2 Protocol-Based Attacks

All protocols run on top of the IP protocol, and the IP protocol has its own set of weakness. For example, DNP3 (distributed network protocol) implements TLS (transport layer security) and SSL (security sockets layer) encryption, which is weak. The protocol is vulnerable to out-of-order, unexpected, or incorrectly formatted packets. Besides the IP protocol, vulnerabilities may exist in smart grid relevant protocols. For example, a significant weakness for IEC 61850 (standard for design of substation automation) is that it maps to manufacturing message specification as the communications platform, which itself has a wide range of potential vulnerabilities. Protocol based attacks must be addressed according to a specific protocol. As mentioned earlier, standards and regulations are required to test any protocols that are proposed to secure smart grid communications infrastructure.

10.5 Existing Security Solutions

General solutions to cybersecurity can be applied to the secure smart grid communications infrastructure. Examples include security by obscurity, requiring a smart grid system to trust no one, applying a layered security framework, or deploy an efficient firewall, intrusion detection systems (IDS), and a self-healing security systems.

The authors in [176] presented a layered specification-based IDS to target ZigBee technology. The proposed design of the IDS is based on anomalous event detection. The work addressed some security issues in the physical and media access control layer. The normal behavior of the network is defined through selected specifications extracted from the IEEE 802.15.4 standard. Deviations from the defined normal behavior are viewed as a sign of malicious activities. The performance analysis demonstrated that the designed IDS provides a good detection capability against both known attacks and unknown attacks. The authors in [7] proposed to use message authentication code (MAC) to authenticate each message and prevent accidental and malicious data corruption en route. Aggregate MAC is often used, since the communication channel capacity is often small and the data size is short compared to the MAC code. However, the aggregate MAC is not resilient against DoS attacks. The authors in [177] applied two security protocols of WLAN (or Wi-Fi) to a smart grid mesh network with a periodic key refreshment strategy. The proposed scheme can achieve simultaneous authentication of equals and efficient mesh security association. The security against DoS attack was improved in this key distribution solution.

Some security solutions are proposed specifically for smart grid communications network and components, especially in the area of private networks [10, 34, 171]. Much of the existing research is focused on AMI, since it is the core of DR in the smart grid. The authors in [178] proposed a privacy-preserving metering system to preserve the privacy of consumers in the smart grid. In the proposed system, a user grants a service provider an access right to meter readings at a time granularity. Meter readings are securely stored in a semitrusted storage system. The authors in [179] proposed a privacy-aware smart metering protocol: smart meter speed dating (SMSD). This protocol uses a peer-to-peer masking technique optimized for a small number of participating smart meters. The advantage of this protocol is its low demands on hardware and communication networks.

Metering data collected in the AMI is undoubtedly large in volume and refreshes frequently [34]. With more deployment of renewable energy sources, a large variety of data will also be introduced to the smart grid, such as ambient environmental status, storage unit status, and weather forecasts. Therefore, big data analytics is expected to become part of the smart grid [133, 180]. Cloud computing has been introduced to the smart grid so that big data analytics can take place [133, 181, 182]. Moreover, ID-based cryptographic schemes have been widely studied [182–185]. Unlike well-known symmetric cryptographic schemes (e.g. advanced encryption scheme), ID-based cryptographic schemes need to be redesigned or modified for different applications in the proposed ICT framework due to various requirements. For instance, some data in our framework requires both confidentiality and non-repudiation while the computation needs to be efficient; some data requires non-repudiation only; the domain secrets need to be refreshed frequently, etc.

A more comprehensive information communications technology framework is required in the smart grid to better evaluate security in the communication infrastructure. For instance, a framework may include private networks set by a utility company, a hybrid cloud-based control center with sensitive data collected and preprocessed at local control centers, and a more visionary idea of harvesting data from various public sources. With that, security can be designed and allowed to evolve as the smart grid evolves.

10.6 Standardization and Regulation

In the past years, many standards and regulations have been proposed for the smart grid communication infrastructure.

10.6.1 Commissions and Considerations

The Energy Independence and Security Act (EISA) of 2007 is a public law to move the United States toward greater energy independence and security; to increase the production of clean renewable fuels; to protect consumers; to increase the efficiency of products, buildings, and vehicles; to promote research on and deployment of greenhouse gas capture and storage options; and to improve the energy performance of the Federal Government, as well as other purposes.

In particular, EISA 2007 directs the NIST to coordinate the development of model standards for interoperability of smart grid devices and systems by 1) creating flexible, uniform, and technology neutral standards and 2) enabling traditional resources, distributed resources, renewables, storage, efficiency, and demand response to contribute to an efficient, reliable grid. Moreover, EISA 2007 directs the Federal Energy Regulatory Commission (FERC), when sufficient consensus exists, to adopt standards necessary to insure smart-grid functionality and interoperability in the interstate transmission of electric power and regional and wholesale electricity markets. However, EISA 2007 did not expand the FERC's Federal Power Act authority to enforce standards.

Regulation may adopt standards separately or in parallel with FERC. State commission may also consider standards when approving utility investments. When adopting standards, regulators need to ensure interoperability and security, without impeding innovation. Regulators also need to consider that consistent action will influence the vendor community. Some vendors often will follow standards that are not legally mandated.

10.6.2 Selected Standards

Table 10.4 lists a few selected standards proposed for the traditional power grid and the smart grid. IEEE Stardard P2030, "Guide for Smart Grid Interoperability of Energy Technology and Information Technology Operation with the Electric Power System (EPS), and End-Use Applications and Loads" provides a knowledge base addressing terminology, characteristics, functional performance and evaluation criteria, and the application of engineering principles to smart grid interoperability of the electric power system with end-use applications and loads [186].

Table 10.4 Selected standards for the Smart Grid.

Institute of Electrical and Electronics Engineers

IEEE Std 2030	Power Engineering Technology
	Information Technology
	Communications Technology

International Electrotechnical Commission

IEC 61968	Distribution Management
IEC 61970	Common Information Model
IEC 60870	Intercontrol Center Communication Protocol
IEC 62351	Data and Communication Security
IEC 62357	Reference Architecture
IEC 61850	Standard for Design of Substation Automation
IEC 61850-7-420	Integration of Distributed Energy Resources
IEC 61850-7-410	Integration of Hydro Resources
IEC 61400	Integration of Wind Farms to Utility Communication Network
IEC 62056	Communication

The International Electrotechnical Commission (IEC) has published over 100 standards that are relevant to the smart grid. In particular, IEC 62351, "Power systems management and associated information exchange—Data and communications security" is relevant to EMS, DMS, DA, SA, DER, AMI, DR, smart home, storage, and EVs in the smart grid. IEC 62351 has seven categories, where each one defines specifications for a certain area.

- IEC/TS 62351-1: Communication network and system security - Introduction to security issues.
- IEC/TS 62351-2: Glossary of terms.
- IEC/TS 62351-3: Profiles including TCP/IP.
- IEC/TS 62351-4: Profiles including MMS.
- IEC/TS 62351-5: Security for IEC 60870-5 and derivatives.
- IEC/TS 62351-6: Security for IEC 61850.

- IEC/TS 62351-7: Network and system management (NSM) data object models.
- IEC/TS 62351-8: Role-based access control.

Other security standards and regulations have been developed for the current power grid and/or the smart grid communications infrastructure in the past. Some examples are:

- DISA Security Technical Implementation Guides (STIGs).
- FIPS 201 (Federal Information Processing Standard Publication 201): a U.S. federal government standard that specifies Personal Identity Verification (PIV) requirements for federal employees and contractors.
- North American Electrical Reliability Corporation-Critical Infrastructure Protection (NERC CIP).
- National Infrastructure Protection Plan (NIPP).
- IEEE 1402: IEEE guide for electric power substation physical and electronic security.
- International Society of Automation(ISA).
- ISO/IEC 17799: Information technology, security techniques, code of practice for information security management.
- Domain Expert Working Groups (DEWGs): Consists of NIST and GridWise Architecture Council (GWAC) to explore smart grid interoperability issues, including:
  - Home-to-Grid;
  - Building-to-Grid;
  - Industrial-to-Grid;
  - Transmission and Distribution;
  - Business and Policy.

Given the continuous evolution of the smart grid and the massive scale as well as complexity of the cyber-physical system, even more standards and regulations are required by the smart grid communications infrastructure, especially for security.

10.7 Summary

In this chapter, we discussed security challenges and some solutions for the smart grid communications infrastructure. The challenges come from the technical requirements of various types of applications in the smart grid. They are also from the continuous operation and evolution of the smart grid. A logical security architecture published by the NIST can be used as a guideline for security planning. Security requirements in smart grid communications network shall be considered depending on applications using private or public network. In additon, although many standards and regulations have been proposed the smart grid, more are required to better guide efforts to secure the smart grid communications infrastructure.

11

Security Schemes for AMI Private Networks

In this chapter, a security protocol is proposed specifically for the advanced metering infrastructure (AMI) in the smart grid to address the security requirements. Although AMI does not cover all private networks in smart grid communications, security protocols for its comprehensive and complicated infrastructure can be extended to other private networks based on their different security requirements. The proposed security protocol will be illustrated in four parts, namely the initial authentication scheme, secure uplink transmission scheme, secure downlink transmission scheme, and domain secret update scheme.

11.1 Preliminaries

The proposed security schemes in this chapter are based on several network security concepts: for example, security services, security mechanisms, etc. To make the illustration clearer, we first present some basic background about network security in this section.

11.1.1 Security Services

Security services are provided in a system design to protect against possible security attacks. In a communication system, possible security services are described in Table 11.1

For a given communication system, whether to apply a security service depends on system requirements. The system may be most secure if all security services are applied. However, due to limited computational resources, it is impractical to implement all services.

Smart Grid Communication Infrastructures: Big Data, Cloud Computing, and Security,
First Edition. Feng Ye, Yi Qian, and Rose Qingyang Hu.
© 2018 John Wiley & Sons Ltd. Published 2018 by John Wiley & Sons Ltd.

Table 11.1 Security services.

Security service	Description
Access control	Control access from authorized users to resources.
Authentication	Verify the identities of entities.
Confidentiality	Ensure that information is accessible only to authorized entities.
Data integrity	Maintain the accuracy and completeness of data.
Non-repudiation	Prove the data origin.
Availability	Make information available to authorized entities when needed.

11.1.2 Security Mechanisms

Security mechanisms are the tools to achieve security services in a communication system. Some major security mechanisms include encipherment, authentication, access control, digital signature, and data integrity.

Encipherment mechanisms mainly provide confidentiality. Encipherment algorithms include reversible and irreversible ones. Reversible encipherment algorithms are widely known as encryption algorithms, which are divided into symmetric and asymmetric ones.

- A symmetric encryption is also known as private-key encryption. It is an encryption algorithm that is based on a preshared secret key. Both communication entities can perform the same function, either encryption or decryption, using the same process. In this chapter, a symmetric encryption algorithm is denoted as $E_K(M)$, where K is the preshared secret key and M is the input message. The decryption algorithm is denoted as $D_K(M)$ correspondingly.
- An asymmetric encryption algorithm is also known as public-key encryption. In an asymmetric algorithm, each communication entity has two keys, public and private respectively. The public keys are shared with the other communication entities, whereas the private keys are kept secret. A message encrypted with the public key must be decrypted by its paired private key. A message signed with the private key must be verified by its paired public key. Therefore, the two communication entities are not symmetric.
- Irreversible encipherment algorithms are not used for encryption. Those algorithms may or may not depend on a key. They are mostly applied to other services, for example, data integrity and digital signatures.

Encipherment mechanisms are widely applied to achieve other security mechanisms.

Authentication mechanisms are applied to authentication services. Instead of specific algorithms, authentication mechanisms are mainly handshake protocols with messages created by other mechanisms using unique information provided by entities.

Access control mechanisms are applied to control access to services. Similar to authentication, many access control mechanisms use handshake protocols to determine and enforce the access rights of an entity depending on its authenticated identity.

Digital signature mechanisms are applied to non-repudiation services and sometimes to data integrity. Most digital signature mechanisms are based on public-key encryption algorithms, where the sender signs a message with the private key and the receiver verifies the signature with the sender's public key.

Data integrity mechanisms are applied to data integrity services. Irreversible encipherment algorithms are applied as data integrity mechanisms in most communication systems. For example, hash functions are data integrity mechanisms. In this chapter, a hash function is defined as $H(\cdot)$. Note that a hash function has no key, so any entity has the ability to generate and verify a hash code with a given message.

Note that security mechanisms are not a one-to-one match with security services. A security service may be provided by multiple security mechanisms, and a security mechanism may be applied to multiple security services.

11.1.3 Notations of the Keys Used in This Chapter

For simplicity, the notations of the keys used in the proposed security protocol of this section are listed in Table 11.2.

11.2 Initial Authentication

11.2.1 An Overview of the Proposed Authentication Process

In this section, we present the *initial authentication* of the security scheme. The security scheme is proposed mainly for the wireless networks in the AMI. Security in the backbone network can be achieved using existing mechanisms and protocols controlled by utilities. Nodes in the wireless networks mainly include data aggregate points (DAPs) and smart meters.

Table 11.2 Notations of the keys.

Keys for symmetric algorithms	
K_i	Preshared secret key for node n_i
k_i	Active secret key for node n_i
$k_{i,j}$	Session key between n_i and n_j

Keys for asymmetric algorithms	
Pu_i	Public key for node n_i
Pr_i	Private key for node n_i
Pu_{AS}	Public key for the authentication server
Pr_{AS}	Private key for the authentication server

11.2.1.1 DAP Authentication Process

Before joining the AMI, a DAP must be authenticated through the initialization process [167]. Assume that utilities have full control of their private networks; thus an authentication server (AS) is managed by utilities on their side. The initial authentication is based on a trusted AS. Generally speaking, if a node is closer to the AS, it will be authenticated before those that are farther away due to multihop networks. Therefore, before a smart meter joins the AMI, the gateway DAPs and normal DAPs are initialized by the AS. Note that gateway DAPs are initialized before normal DAPs since they have direct communication to the concentrator. For simplicity, gateway DAPs are not specified in the rest of the discussion.

DAPs are divided into two groups: one *active* and the other *uninitialized*. An active node is a node that has been authenticated by the AS to join the AMI communication system and is functioning normally. An uninitialized node can be one of the following four types:

- A newly installed node
- A node that has recovered from a malfunction
- A node updated with new preshared keys
- A node reinstalled to another location

The authentication process for uninitialized nodes is carried by existing active nodes and the AS. For example, as illustrated in Figure 11.1, DAP n_1 is uninitialized, while its neighboring DAPs n_2,

Figure 11.1 Initial authentication process for DAP n_1.

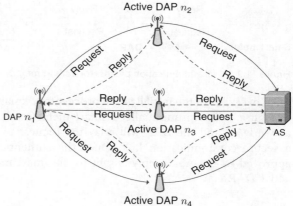

Active DAP n_2

DAP n_1

Active DAP n_3

Active DAP n_4

AS

n_3 and n_4 are active. To join the AMI, n_1 initiates the authentication process by broadcasting a request to all of its active neighbors, which will relay the request to the AS through established secure links.

After the authentication process is performed by the AS, n_1 will receive reply messages from the AS through its active neighbors. The reply messages are different from each other, since they consist of authentication confirmation as well as information to establish secure links between n_1 and each of its active neighbors. In summary, the initial authentication process accomplishes the following three tasks:

- n_1 is authenticated to become an active node and join the AMI
- n_1 establishes a secure connection to the AS through one of its active neighbors that has the shortest distance to the AS
- n_1 establishes backup secure connections to the AS through the rest of its active neighbors

11.2.1.2 Smart Meter Authentication Process

After all DAPs have been initialized by the AS, a neighborhood-area network is formed in the AMI. Smart meters will then be initialized through active DAPs. Unlike DAPs, smart meters do not have many neighbor nodes because of two reasons. First, smart meters have a limited transmission range. They are unlikely to have a direct connection with more than one DAP. Second, it is not a good idea to let smart meters communicate with each other, since their data contains much private information and smart meters are easier to access than DAPs.

Figure 11.2 Initial authentication process for a smart meter.

An overview of the initialization process for a smart meter is shown in Figure 11.2. An uninitialized smart meter sends a request to an active DAP, and the DAP will relay the request to the AS through a secure communication link. Once the authentication process is approved by the AS, a reply is sent to the smart meter through the active DAP.

11.2.2 The Authentication Handshake Protocol

Without loss of generality, the authentication handshake protocol is presented according to the example given in Figure 11.1, where n_1 is the uninitialized node. One of its active neighbor n_2 is chosen to illustrate the detailed initialization process. The initialization processes through active neighbors n_3 and n_4 are similar.

In practice, the active neighboring node n_2 may not have a direct connection to the AS. Instead, there may be a secure link established between n_2 and the AS. Therefore, we focus on n_1, n_2, and the AS in the process. Note that other active nodes in the same secure link do not store useful information from the process. The initialization process achieves three mutual authentications.

- Authentication between n_1 and the AS: This mutual authentication is straightforward, since the AS will only allow legitimate nodes to join the AMI and the nodes will also trust only the AS.
- Authentication between n_2 and the AS: This mutual authentication is intended to ensure that n_2 is active and is trusted to relay the request from n_1.
- Authentication between n_1 and n_2: This mutual authentication between n_1 and n_2 is to help establish further secure communications from n_1 to n_2.

Each legitimate node, whether uninitialized or active, has a pre-shared secret key (i.e. K_i for node n_i) with the AS. Besides, an *active secret key* is assigned to each active node (e.g. key k_i) for node n_i, mainly for uplink data traffic encryption. This active secret key is also used by the AS to verify if this node is active or not. Similar to K_2, k_2 is known only to n_2 and the AS. In order to establish a secure connection from n_1 to the AS after the authentication process, an active secret key k_1 must be generated by the AS and assigned to n_1 during the initialization process. Note that n_1 does not carry k_1 before initialization process; only K_1 is known to n_1. Moreover, although the active nodes relay messages from the AS to n_1, the keys are not disclosed to those intermediate nodes.

The authentication handshake protocol is initiated from the uninitialized node n_1. The entire process includes six handshakes among n_1, n_2, and the AS, as shown in Figure 11.3. The 6 messages exchanged during the process are described as follows:

1) $M_1 = \text{request} \| ID_1 \| t_1 \| H(ID_1 + K_1 + t_1)$.
   To initiate the process, n_1 sends M_1 to the AS through n_2. In message M_1, $H(\cdot)$ is a cryptographic hash function, '+' is an XOR function, and t_1 is a time stamp. The authentication is achieved by K_1, since with given ID_1 and t_1, the AS is the only entity other than n_1 able to compute $H(ID_1 + K_1 + t_1)$.

2) $M_2 = ID_2 \| E_{k_1}(M_1 \| ID_2 \| t_2) \| H(ID_2 + k_2 + t_2)$.
   Once n_2 receives the request message M_1, it generates M_2 and sends the new message to the AS. In M_2, $E_k(\cdot)$ is a symmetric encryption function with key k. In particular, the active key for n_2 is applied to the encryption. Another time stamp t_2 is generated at n_2 and included in M_2 together with its own identity ID_2. Then, n_2 encrypts

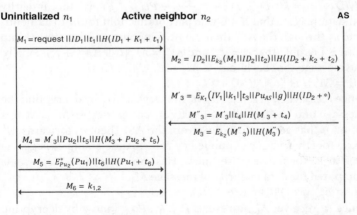

Figure 11.3 Detailed initial authentication process through one active neighbor.

the entire message with k_2. The encryption protects M_1 from being disclosed to other active nodes in the secure link. An identity verification code $H(ID_2 + k_2 + t_2)$ is generated at n_2. The extra information generated at n_2 is used for the AS to authenticate n_2 as an active node and validates the freshness of the message.

3) $M_3 = E_{k_2}(M_3'')\|H(M_3'')$.

Once the AS receives M_2, it authenticates n_2 by decrypting M_2 using k_2. Time stamp t_2 is verified by comparing a computed $H(ID_2 + k_2 + t_2)$ to the received value. Then, the AS authenticates n_1 by computing $H(ID_1 + K_1)$. Once n_1 is authenticated, the AS generates a message $M_3' = E_{K_1}(IV_1\|k_1\|t_3\|PU_{AS}\|g)\|H(ID_2+ *)$ for n_1. In M_3', IV_1 is the initial vector for further uplink transmission, and k_1 is the active key for n_1. Pu_{AS} is the public key of the AS for downlink transmission security protocols. Moreover, g is the generating parameter for public-key cryptography in the communication domain. Parameter g can be a set of parameters depending on the chosen public-key cryptographic schemes. For example, g can be two primes numbers if RSA [187] is applied and have more parameters if identity-based cryptography [188, 184, 185] is applied. Nonetheless, g remains the same in the communication domain. Although the AS generates g, it does not generate public/private keys for each node. It is better to keep the nodes as independent as possible from other nodes and the AS. Unique information for n_1 is encrypted with the preshared secret key K_1, that is, $E_{K_1}(IV_1\|k_1\|t_3\|PU_{AS}\|g)$. Moreover, in M_3', $H(ID_2+ *) = H(ID_2 + IV_s + k_1 + t_3 + PU_{AS} + g)$ is for integrity validation. Note that ID_2 is also part of the input for n_1 to authenticate n_2 through the AS. Then the AS generates another time stamp t_4 (it is possible that $t_4 = t_3$) and $M_3'' = M_3'\|t_4\|H(M_3' + t_4)$. Finally, the message sent back to n_2 is $M_3 = E_{k_2}(M_3'')\|H(M_3'')$.

4) $M_4 = M_3'\|Pu_2\|t_5\|H(M_3' + Pu_2 + t_5)$.

Once message M_3 reaches n_2, n_2 first reveals M_3'' by decrypting the message and verifies data integrity by computing $H(M_3'')$. At this point, n_2 has authenticated n_1 from the AS. Then, n_2 forwards M_3' to n_1 together with the public key Pu_2. A time stamp t_5 is generated at n_2 for message freshness. Hash value $H(M_3' + Pu_2 + t_5)$ is computed for data integrity of message M_4.

5) $M_5 = E_{Pu_2}^*(Pu_1)\|t_6\|H(Pu_1 + t_6)$.

Once n_1 receives M_4, it reveals IV_1, k_1, Pu_{AS}, and g by decrypting M_3'. After verifying the integrity of the received information,

n_1 successfully authenticates n_2. Then n_1 generates a pair of public/private keys based on a given public domain secret g. The public key Pu_1 is encrypted with the public key of n_2 so that $E^*_{Pu_2}(Pu_1)$, where E^* is the encryption function of the adopted public key cryptography. A time stamp t_6 is generated, and $H(Pu_1 + t_6)$ is computed at n_1 to provide data integrity for message M_5.

6) $M_6 = k_{1,2}$.

After exchanging public keys Pu_1 and Pu_2, active nodes n_1 and n_2 can find a way to generate the session key $k_{1,2}$ for further communication. Session key $k_{1,2}$ is shared only between n_1 and n_2. It is refreshed frequently.

After the six handshakes, n_1 is fully initialized, and it joins AMI communications through n_2. The initial authentication processes through other active neighbors (i.e. n_3 and n_4 in this example) are similar. In particular, the AS sends back the same IV_1, k_1, Pu_{AS}, and g. Note that the preshared secret key $k_{1,x}$ is unique, depending on active neighbor identity x. Nonetheless, in the final handshake, n_1 will send the same Pu_1 to its active neighboring node n_x encrypted with Pu_x, $x = 2, 3, 4$ in this example. By doing so, n_1 shares the same public key to all of its active neighbors. Therefore, n_1 is able to join the uplink transmission through any of the active neighbors. In other words, both operating and backup secure communication channels are established through the initial authentication process.

The detailed process for smart meter initialization is the same as DAP initialization. The only difference is that, there is one process only because of the single active DAP a smart meter is connected to. The illustration is not repeated here.

11.2.3 Security Analysis

The security of the proposed authentication process is described in terms of each security service, including confidentiality, data integrity, non-repudiation, and availability. Note that the security services mentioned here are just those for the authentication process, not for the overall security protocol.

- **Confidentiality.** Confidentiality of the authentication request is not necessary; therefore, it is not provided via specific mechanisms. Much information is transmitted in clear text. Note that the initial authentication request is protected with confidentiality once it reaches an active node in the AMI.

- **Data integrity.** All the messages (except for M_6) are provided a hash value specifically for an integrity check. Moreover, the input is not the original message, which can easily be captured by an eavesdropper. The input is the XORed messages of the useful information, which cannot be captured or forged. Therefore, the messages in this protocol is cannot be forged. Moreover, with time stamps being applied in each message, a replay attack is unlikely to succeed in the process. The detailed process of M_6 is not given in this protocol, because the real application may vary based on different public key cryptographic schemes. With a given public key cryptographic scheme, data integrity can be provided in a similar way for session key $k_{i,j}$ in M_6.
- **Non-repudiation.** The initialization process does not use a digital signature for sender authentication except for M_5. However, secret preshared keys are applied for message encryption. With the sender and the receiver being the only entities that can encrypt and decrypt the message, non-repudiation is achieved for all messages (except for M_5) with symmetric encryption. Non-repudiation of M_5 is indeed provided by a digital signature.
- **Availability.** Availability of the process is guaranteed as long as the AMI has a wireless connection. The proposed authentication process has enhanced availability through the participation of all active nodes that are neighbors of an uninitialized one.

11.3 Proposed Security Protocol in Uplink Transmissions

In the uplink transmission, data from each node is aggregated in a chain topology and is finally delivered to the service provider (assuming that the AS and the service provider share the same entity). As discussed before, data confidentiality and data integrity are critical requirements for metering data, since any mistakes may cause inefficient grid operation. Sender authentication or non-repudiation may be considered in certain situations if there are enough computational resources. To achieve all these requirements, we propose the following security protocol for data aggregation in uplink transmission.

11.3.1 Single-Traffic Uplink Encryption

We first present the uplink encryption process for a single-traffic transmission. The following description is based on the illustration shown in Figure 11.4. Suppose the single-traffic transmission follows a path with N nodes in the order of $(n_1, n_2, \ldots, n_N)$.

As the first one of the aggregation, n_1 mixes its raw data D_1 with IV_1 as $D_1 \oplus IV_1$. It then encrypts the intermediate message with the active secret key k_1 as follows:

$$C_1 = E_{k_1}(D_1 \oplus IV_1). \tag{11.1}$$

A hash value is generated as $H^*(C_1)$, where $H^*(\cdot)$ is a hashed message authentication code function that provides data integrity. The notation with $*$ is to distinguish the hash function from the one applied to the initialization process, since different hash functions can be used in authentication and uplink encryption. The hash value is attached to the cipher text as follows:

$$M_1 = C_1 \| H^*(C_1). \tag{11.2}$$

Finally, n_1 encrypts the entire message with $k_{1,2}$ as follows:

$$M_{1,2} = E_{k_{1,2}}(M_1). \tag{11.3}$$

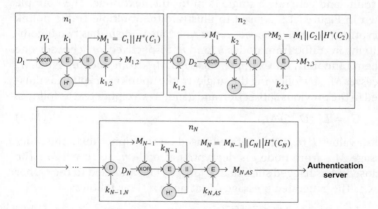

Figure 11.4 Data aggregation process in an uplink transmission.

The next node n_2 first decrypts the incoming data with the session key as follows:

$$M_1 = D_{k_{1,2}}(M_{1,2}). \tag{11.4}$$

Then n_2 mixes its raw data D_2 with M_1 and generates cipher text C_2 as follows:

$$C_2 = E_{k_2}(D_2 \oplus M_1). \tag{11.5}$$

A hash value is generated as $H^*(C_2)$ and attached to the cipher text. The message from n_1 is aggregated to the current message as follows:

$$M_2 = M_1 \| C_2 \| H^*(C_2). \tag{11.6}$$

Finally, n_2 encrypts the entire message with $k_{2,3}$ as follows:

$$M_{2,3} = E_{k_{2,3}}(M_1). \tag{11.7}$$

Any intermediate node n_i performs the same process as n_2 by replacing indexes $1, 2$ with $i-1, i$ respectively. The final message reaching the AS (or utilities) from node n_N is $M_N = M_{N-1} \| C_N \| H^*(C_N)$.

11.3.2 Multiple-Traffic Uplink Encryption

A node may receive multiple items of data traffic in the AMI communications network. In this case, the intermediate shall process all incoming traffic and generate a single item for the next node. The example shown in Figure 11.5 is used to illustrate the multiple-traffic uplink encryption process. Suppose an intermediate node n_p has two incoming items in traffic from nodes n_i and n_j respectively; n_p chooses one of them randomly, for example, message $M_{i,p}$ from node n_i. Node n_p processes $M_{i,p}$ by following the single-traffic uplink encryption as illustrated in the previous subsection and generates a cipher text as follows:

$$C_p = E_{k_p}(D_p \oplus M_i). \tag{11.8}$$

A hash value $H^*(C_p)$ is generated based on the cipher text. The other message $M_{j,p}$ from node j is decrypted at node n_p to reveal M_j. The disclosed M_j is flagged such that $f_0 \| M_j \| f_1$ and attached to the cipher text C_p. The generated message M_p at node n_p is as follows:

$$M_p = f_0 \| M_j \| M_i \| C_p \| H^*(C_p). \tag{11.9}$$

The message sent from n_p to the next node is computed as follows:

$$M_{p,p+1} = E_{k_{p,p+1}}(M_p). \tag{11.10}$$

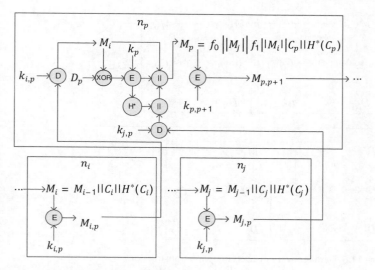

Figure 11.5 Multiflow data aggregation process.

If there are more items in incoming traffic, they will be processed as $M_{j,p}$ at node n_p.

11.3.3 Decryption Process in Uplink Transmissions

Once the AS receives the aggregated data, it starts the decryption process of the data. For example, as shown in Figure 11.6, the AS first authenticates the incoming node, for example, node n_N, by decrypting the receiving data with the preshared public key $Pu_{N,AS}$ as follows:

$$M_N = D_{k_{N-1,AS}}(M_{N,AS}) \tag{11.11}$$

and reveals C_N, H^*C_N and M_{N-1}. The AS then verifies the data integrity by computing the hash value H^*C_N. As shown in Figure 11.7, if the hash value matches the one disclosed from the decryption, then data integrity is validated. If a hash value cannot be validated, then the data integrity of that message is violated, and the message is discarded. Once data integrity is validated, the AS continue the data recovery process by decrypting C_N as follows:

$$D_N \oplus M_{N-1} = D_{k_N}(C_N). \tag{11.12}$$

With M_{N-1} already disclosed, D_N from node n_N can be recovered. After that, the AS continues the recovery process with M_{N-1}. Note

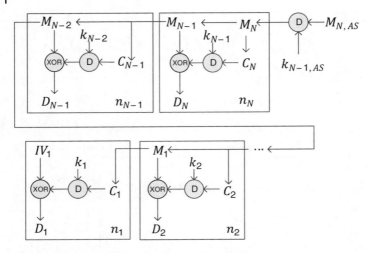

Figure 11.6 Data recovery process in uplink transmission.

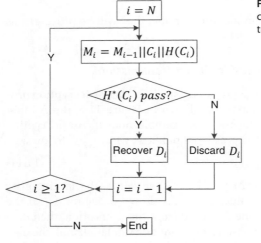

Figure 11.7 Data integrity check in uplink transmission.

that the AS does not verify senders other than n_N; thus M_{N-1} is processed using the same process used to process M_N. To recover D_1, the AS maintains synchronization of the initial vector IV_1.

If the message includes data from multiple incoming traffic streams, the AS extracts the messages between flags f_0 and f_1 and continues to recover the data by following the same process discussed before for D_N. The AS may process multiple incoming messages in parallel, as they are not mixed together.

11.3.4 Security Analysis

Confidentiality, data integrity, and non-repudiation are the security services to be provided in the uplink transmission encryption protocol.

- **Confidentiality.** Confidentiality is achieved in two steps. First, the raw data D_i from node n_i is mixed with the incoming data from the previous node, that is, M_{i-1}. The first node achieves this step by mixing its data with the initial vector given by the AS. In addition, the mixed data is encrypted with the active key k_i.
- **Data integrity.** The data integrity of each message is provided by a hash value generated from that message. The message M_i from of n_i is cannot be forged unless its active key k_i is compromised.
- **Non-repudiation.** Non-repudiation is provided in two aspects. Within one-hop transmission, a message $M_{i,i+1}$ is encrypted by the preshared secret key $k_{i,i+1}$ that is known only to nodes n_i and n_{i+1}. Thus non-repudiation is provided to sender n_i. At the AS, messages M_i for all nodes are encrypted with their active key k_i; thus sender authentication is provided.

11.4 Proposed Security Protocol in Downlink Transmissions

The downlink transmission involves control messages from the service provider to the nodes. Most of the control messages (e.g. price and tariff information) are for all the smart meters in the neighborhood. For those messages, confidentiality may not be as important as it is for the uplink transmission data. Nonetheless, data integrity is important. Message manipulation would alter demand response in the smart grid and in the end result in grid inefficiency. Moreover, non-repudiation is critical for such control messages so that the customers can trust the sender. Some control messages are one-to-one, such as on-off switch commands for participating customers' air conditioners. Confidentiality must be provided to those messages in addition to data integrity and non-repudiation.

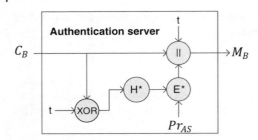

Figure 11.8 Encryption of broadcast control message M_B.

11.4.1 Broadcast Control Message Encryption

Encryption for a broadcast control message (e.g. C_B) is illustrated in Figure 11.8. A time stamp t is appended to the message, and a hash value is generated as $H^*(C_B \oplus t)$, where $H^*(\cdot)$ is a cryptographic hash function. The hash value is signed by the AS using its private key, such that $E^*_{Pr_{AS}}(H^*(C_B \oplus t))$, where $E^*_{Pr_{AS}}(\cdot)$ is an encryption function using public key cryptography. The broadcast message to each node includes the original control message, time stamp, and the digital signature, such that,

$$M_B = C_B \| t \| E^*_{Pr_{AS}}(H^*(C_B \oplus t)). \tag{11.13}$$

At the receiver side, the original information (i.e. C_B and t) is in clear text. The digital signature is decrypted using the public-key of the AS by performing $D^*_{Pu_{AS}}(M_B)$, where $D^*_{Pr_{AS}}(\cdot)$ is a decryption function using public key cryptography. The decrypted information is the hash value. The receiver shall compute the hash value at its side and compare the result with the decrypted value to verify data integrity. If the hash value is valid, then non-repudiation is also validated. If the integrity check is not passed, the receiver will request a retransmission from the AS through its secure uplink transmission tunnel. This rarely happens unless the message is not legitimate. Because each node will receive multiple copies of the control message from all of its active neighbors, if one of the messages is valid, then a retransmission will not be necessary.

11.4.2 One-to-One Control Message Encryption

Encryption for a one-to-one control message (e.g. C_i) is illustrated in Figure 11.9. The message is XORed to a time stamp t and encrypted using the active key k_i. A hash value of the encrypted data is generated

Figure 11.9
Encryption of
control message
M_i for n_i.

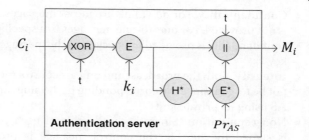

and signed. Finally, the encrypted data, time stamp, and the digital signature are aggregated together as message M_i such that

$$M_i = E_{k_i}(C_i \oplus t)\|t\|E^*_{Pr_{AS}}(H^*(C_i \oplus t)). \tag{11.14}$$

Unlike M_B, M_i is sent through all of its active neighbors of n_i only, as illustrated in Figure 11.10. A few copies of the information would increase the reliability of transmissions.

11.4.3 Security Analysis

Security services provided to broadcast messages and one-to-one messages in downlink transmissions are slightly different.

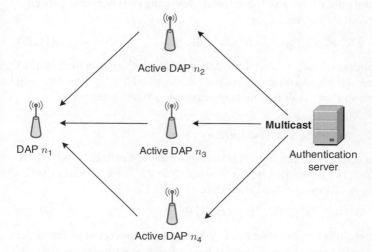

Figure 11.10 Example of control message M_1 to n_1.

- **Confidentiality.** For downlink broadcast messages, confidentiality is not provided. For one-to-one messages to a specific node (e.g. n_i), confidentiality is provided by encrypting the message with the active key k_i.
- **Integrity.** Both the broadcast and unicast control messages are cannot be forged, since the corresponding hash values are signed by the AS using its private key.
- **Non-repudiation.** Since the hash value of each control message is signed by the AS, the control message is protected against repudiation.

11.5 Domain Secrets Update

In order to keep the AMI secure in the long run, domain secrets need to be refreshed frequently, for example, daily or hourly depending on requirements.

11.5.1 AS Public/Private Keys Update

For the AS, its public and private keys need to be refreshed. After the AS generates a new pair of public/private keys (i.e. Pu'_{AS}/Pr'_{AS}), it transmits the public key to all the active nodes using the encryption scheme for broadcasting, such that

$$M_B = Pu'_{AS}\|t\|E^*_{Pr_{AS}}(H^*(Pu'_{AS} \oplus t)). \tag{11.15}$$

The update of Pu'_{AS} is for all the active nodes in the same time slot. In the mean time, separate control messages signed by Pr_{AS} and Pr'_{AS} will be sent so that the downlink transmissions are not interrupted.

11.5.2 Active Secret Key Update

For an active node (e.g. n_i), its active secret key k_i needs to be refreshed. To do so, the AS picks a new active secret key k'_i for n_i, and sends it as one-to-one message to n_i, such that

$$M_i = E_{k_i}(k'_i \oplus t)\|t\|E^*_{Pr_{AS}}(H^*(k'_i \oplus t)). \tag{11.16}$$

However, it is not necessary to refresh the active secret keys for all the nodes at the same time. The AS can do a batch at a time when the network is not heavily loaded, for example, after midnight. Moreover, as mentioned before, the session key (e.g. $k_{i,j}$) between two active nodes (n_i and n_{i+1}) needs to be refreshed more frequently. To do so, n_i and n_j simply run the sixth step from the initialization process again.

11.5.3 Preshared Secret Key Update

The preshared key of a node is not refreshed as frequently as the other keys since it is used much less frequently. Therefore, the preshared key can last longer before it wears out. However, it is reasonable to refresh the preshared key in some cases, for example, if a DAP is compromised and recovered, if a DAP is redeployed to another NAN, or if a house has been sold and thus its smart meter has a new owner. An on-site firmware update will be recommended in this case. A customer can also request a firmware update and then load it to his/her smart meter. Automatic updates can also be achieved. For example, if DAP n_i needs a preshared key update, the AS picks a new K_i', and sends it to n_i, such that

$$M_i = E_{k_i}(K_i' \oplus t)C_B \| t \| E^*_{Pr_{AS}}(H^*(K_i' \oplus t)). \qquad (11.17)$$

It is also reasonable to encrypt this message with K_i if k_i has been compromised. However, if both K_i and k_i are compromised, then a physical update will be inevitable.

11.6 Summary

In this chapter, we proposed a security protocol for the AMI in the smart grid. In order to meet various security requirements for asymmetric communication in the AMI, the proposed security protocol consists of initial authentication scheme, independent security schemes for uplink and downlink transmissions, and a domain secret update scheme. The security scheme for uplink transmissions provides confidentiality and data integrity to metering data and other monitoring data. The security scheme for downlink transmissions provides data integrity and non-repudiation for controlling data and pricing/tariff information. Future work may be conducted to extend the proposed network security protocol so that cloud computing and various external information sources can be involved in the modern control of the smart grid.

12

Security Schemes for Smart Grid Communications over Public Networks

In this chapter, we focus on security schemes for smart grid communications over public networks that utility companies subscribe to. Public networks in the smart grid are distinguished from private networks by their independence from utilities. Our proposed security schemes can be applied by utilities even without full control of the communications network to add extra protection on top of existing security provided by network service providers.

12.1   Overview of the Proposed Security Schemes

12.1.1   Background and Motivation

With the introduction of cloud computing, some data must be transmitted through public networks (e.g. the Internet) in smart grid communications [181, 189, 190]. Although public cloud service providers have certain security mechanisms within the cloud, data exchange over the Internet still needs extra protection. We propose to apply identity-based (ID-based) security schemes [183–185] to secure data transmission over the Internet. With the proposed security schemes, utility companies can have more security control. Moreover, the communication infrastructure in the smart grid can better handle large numbers of participants.

The foundation of an ID-based security scheme is public-key cryptography. Instead of generating keys randomly, an ID-based security scheme utilizes the unique ID of each participant. By doing so, key management might be more convenient, since some of the keys can be computed locally or even ahead of time. Furthermore,

Smart Grid Communication Infrastructures: Big Data, Cloud Computing, and Security,
First Edition. Feng Ye, Yi Qian, and Rose Qingyang Hu.
© 2018 John Wiley & Sons Ltd. Published 2018 by John Wiley & Sons Ltd.

privacy and authentication can still be provided to the participants. In the smart grid information and communication networks (ICT) framework, each component has a unique ID that can be applied to ID-based security schemes. Specifically, the proposed ID-based security scheme utilizes public key cryptography where the public key is computed mainly based on the ID of each participant together with an expiration indicator *time*. Public keys can be computed locally by any legitimate user in the domain. As a result, public keys and related domain secrets can be refreshed easily after each session. Key management is simplified by adopting the proposed ID-based security scheme. With carefully chosen bilinear pairing operation and other system parameters, ID-based security schemes can perform efficiently in smart grid communication systems.

Despite its simplicity, the proposed ID-based based security scheme can provide security services such as confidentiality, data integrity, and non-repudiation to the smart grid communications network. In this chapter, the proposed ID-based security scheme is designed to achieve digital signature and encryption simultaneously; thus the proposed scheme is named ID-based signcryption (IBSC). The IBSC scheme can be reduced to an ID-based digital signature for those cases that do not require confidentiality. In order to enhance performance, the proposed IBSC is also modified for session key distribution instead of direct message encryption. In addition, the ID-based schemes are also applied to achieve delegation of signing rights. With this feature, a utility control center may hand over its data control authority to another (or a few other) control center temporarily during routine maintenance, system failure, etc.

In our proposed ID-based security, we adopt ($ID\|time$) instead of ID for public key generation, where *time* is the expiration time of the current session. Once a session expires, all participants will update the corresponding secrets and parameters accordingly. When a participant leaves the system domain, secrets bared by this participant need to be revoked. By adopting $ID\|time$, if the public key generator (PKG) stops issuing secret keys to the participant that left, key revocation can be done automatically at the beginning of the next session. New messages will not be disclosed to old keys.

12.1.2 Applications of the Proposed Security Schemes in the Smart Grid

The proposed ID-based security schemes can be applied to a variety of applications in the smart grid communication infrastructure. For communications between a utility's local control centers and a cloud control center, the proposed ID-based security schemes may function as follows:

- **Encipherment and digital signature.** The proposed security schemes can be applied directly to provide both confidentiality and non-repudiation. For instance, preprocessed metering data sent from local control centers to a cloud control center is encrypted and signed to provide confidentiality and non-repudiation. Information generated by big data analytics is also encrypted and signed before being sent from the cloud to local control centers.
- **Session key distribution.** If symmetric ciphers are preferred in some applications in the smart grid communication infrastructure, the proposed identity-based scheme can be applied to achieve secure session key distribution.
- **Signing rights delegation from a local control center to another one.** If local control center L_i is subject to routine maintenance, it may delegate signing rights to another local control center (e.g. L_j). As shown in Figure 12.1, the private key generator (PKG) controlled by a utility has the authority to delegate signing rights from L_i to L_j. Alternatively, L_i can delegate signing rights to L_j locally without involving another entity for more efficient operation.

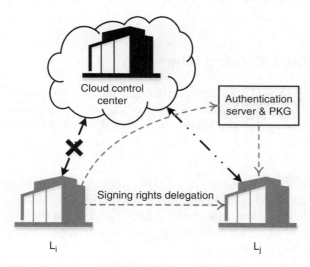

Figure 12.1 Signing rights delegation from L_i to L_j.

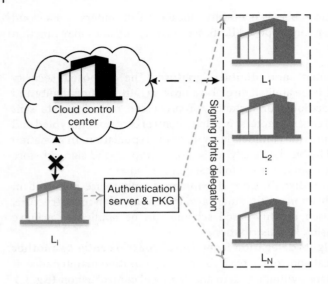

Figure 12.2 Signing rights delegation from L_i to a group of *LCC*s.

- **Signing rights delegation from one local control center to a group of others.** The PKG can assign a group of local control centers as a group proxy to sign for L_i, as illustrated in Figure 12.2. In this case, no other local control center will take full responsibility for L_i.

12.2 Proposed ID-Based Scheme

The core of the proposed solution is an IBSC scheme, which performs the functions of both digital signature and encryption simultaneously. The scheme can be further applied to achieve digital signature only or key distribution.

12.2.1 Preliminaries

The proposed ID-based security scheme is based on a bilinear map. Let $\mathbb{G}_1$ and $\mathbb{G}_2$ be groups of prime order q. Let g be a generator of $\mathbb{G}_1$. Let $\hat{e} : \mathbb{G}_1 \times \mathbb{G}_1 \rightarrow \mathbb{G}_2$. We say that $(\mathbb{G}_1, \mathbb{G}_2)$ are bilinear map groups if $\hat{e}$ has the properties as follows:

- **Bilinearity**. $\hat{e}(aP, bQ) = \hat{e}(P, Q)^{ab}$ for all $P, Q \in \mathbb{G}_1$ and all $a, b \in \mathbb{Z}_q^*$.
- **Nondegeneracy**. For any $P \in \mathbb{G}_1$, $\hat{e}(P, Q) \neq 1$ for all $Q \in \mathbb{G}_1 \setminus \{\mathcal{O}\}$ (indicated as $\mathbb{G}_1^*$ hereafter).
- **Computability**. There is a polynomial time algorithm for computing $\hat{e}(P, Q)$ for all $P, Q \in \mathbb{G}_1$.

12.2.2 Identity-Based Signcryption

The proposed IBSC scheme comprises five algorithms: *Setup*, *Keygen*, *Signcryption*, *Decryption*, and *Verification*. Without loss of generality, the algorithms are described using the case where A ($ID_A = A$) sends message $M = \{0, 1\}^n$ to B ($ID_B = B$).

12.2.2.1 Setup

Setup is the algorithm used to initialize a domain's public parameters and set the public/private keys of the authentication server (AS). For simplicity, the AS and PKG are considered interchangeably in the discussion. In practice, they shall be deployed and maintained separately. In *Setup*, the PKG chooses groups $(\mathbb{G}_1, \mathbb{G}_2)$ of prime order q, a generator g of $\mathbb{G}_1$, a randomly chosen master key $s \xleftarrow{R} \mathbb{Z}_q^*$, and a domain secret $g_1 = sg \in \mathbb{G}_1$. The PKG also chooses three cryptographic hash functions as follows:

$$H_1 : \{0, 1\}^* \to \mathbb{G}_1^*,$$
$$H_2 : \{0, 1\}^* \to \mathbb{Z}_q^*,$$
$$H_3 : \{0, 1\}^* \to \{0, 1\}^n.$$

The domain public parameters are

$$params = \langle \mathbb{G}_1, \mathbb{G}_2, g, q, g_1, H_1, H_2, H_3, n \rangle.$$

The public/private keys of the AS are $p_{AS} = H_1(AS \| time)$ and $d_{AS} = sp_{AS}$ respectively.

12.2.2.2 Keygen

Keygen is the algorithm used to generate public and private keys for each entity in the system. For a given string $ID \in \{0, 1\}^*$ and a expiration time stamp *time*, the algorithm builds a public/private key pair p_{ID}/d_{ID} as follows:

- Public key: $p_{ID} = H_1(ID \| time)$.
- Private key: $d_{ID} = sp_{ID}$.

For example, the keys for A are $p_A = H_1(A\|time)$ and $d_A = sp_A$. Note that *time* is converted into $\{0,1\}^*$ and is concatenated to *ID* in the illustration. Other processes can be taken for the same purpose; for example, *time* can also be XORed to *ID*.

12.2.2.3 Signcryption

Keygen is the algorithm used to encrypt and sign (signcrypt) a message. To signcrypt a message M, sender A

1) randomly picks $r \xleftarrow{R} Z_q^*$ and computes

$$U = rg;$$

2) computes $h_1 = H_2(M\|A\|U)$ and computes

$$V = d_A h_1 + rg_1;$$

3) computes $p_B = H_1(B\|time)$ and $h_2 = H_2(A\|B)$ and then computes

$$X = h_2 U;$$

4) computes $h_3 = H_3(X\|\hat{e}(rg_1, h_2 p_B))$ and encrypts the message

$$W = M \oplus h_3;$$

5) finally outputs a 4-tuple $\langle U, V, X, W \rangle$.

In the 4-tuple, the cipher text is $C = \langle X, W \rangle$ and the digital signature is $\sigma = \langle U, V \rangle$.

12.2.2.4 Decryption

Decryption is the algorithm used to decrypt a cipher text $C = \langle X, W \rangle$. Upon receiving $\langle \sigma, C \rangle$, receiver B decrypts M in the following steps:

1) computes $h_3' = H_3(X\|\hat{e}(X, d_B))$;
2) decrypts $M = W \oplus h_3'$.

12.2.2.5 Verification

Verification is the algorithm used to validate a digital signature $\sigma = \langle U, V \rangle$. Note that the original message M must be recovered before verification. A digital signature is verified by B in the following steps:

1) computes $p_A = H_1(A\|time)$, and $h_1 = H_2(M\|A\|U)$;
2) verifies if $\hat{e}(g, V) = \hat{e}(g_1, p_A h_1 + U)$.

From the illustration we can see that sender A encrypts the message with p_B so that confidentiality is provided. Sender A also signs the message with d_A so that non-repudiation is provided. Data integrity is also provided with hash functions.

12.2.3 Consistency of the Proposed IBSC Scheme

We then verify the consistency of the proposed IBSC scheme, in particular, the algorithms *Decryption* and *Verification*. The original message M can be recovered with algorithm *Decryption* if and only if $h_3' = h_3$. The consistency can be proved as follows:

$$
\begin{aligned}
\hat{e}(X, d_B) &= \hat{e}(h_2 rg, sp_B) \\
&= e(rg, p_B)^{sh_2} \\
&= e(rg_1, h_2\, p_B).
\end{aligned}
\tag{12.1}
$$

Therefore

$$
\begin{aligned}
h_3' &= H_3(X \| \hat{e}(X, d_B)) \\
&= H_3(X \| \hat{e}(h_2 rg, sp_B)) \\
&= h_3.
\end{aligned}
\tag{12.2}
$$

The consistency of algorithm *Verification* is proved as follows:

$$
\begin{aligned}
\hat{e}(g, V) &= \hat{e}(g, d_A h_1 + rg_1) \\
&= \hat{e}(g, p_A h_1 + rg)^s \\
&= \hat{e}(g_1, p_A h_1 + U).
\end{aligned}
\tag{12.3}
$$

12.2.4 Identity-Based Signature

As discussed earlier, not all messages need encryption in smart grid communications. Nonetheless, data integrity and non-repudiation are still required for most messages. To simplify the computation of each node, the IBSC scheme may be reduced to an identity-based signature (IBS) scheme for the purpose of digital signature only. The IBS scheme comprises four algorithms, *Setup*, *Keygen*, *Signature*, and *Verification*. The algorithms *Setup* and *Keygen* are the same as the ones in IBSC. The algorithms *Signature* and *Verification* are described in the following sections. For consistency, assume A sends M to B in the discussion.

12.2.4.1 Signature

Signature is the algorithm in IBS used to sign a message by sender A. For a given message M, sender A signs it in the following steps:

1) randomly picks $r \xleftarrow{R} Z_q^*$ and computes

$$U = rg;$$

2) computes $h_1 = H_1(M\|A\|U) \in \mathbb{Z}_q^*$ and computes

$$V = d_A h_1 + rg_1;$$

3) finally outputs $\sigma = \langle U, V \rangle$.

12.2.4.2 Verification

Verification is the algorithm on the receiver side used to validate a digital signature. The receiver B verifies a digital signature $\sigma = \langle U, V \rangle$ in the following steps:

1) computes $p_A = H_1(A\|time)$ and $h_1 = H_1(M\|A\|U)$;
2) verifies if $\hat{e}(g, V) = \hat{e}(g_1, p_A h_1 + U)$.

This completes the description of the IBS scheme. The consistency has been is proven by Eq. (12.3).

12.2.5 Key Distribution and Symmetrical Cryptography

Although encryption is achieved in the IBSC scheme, some may still prefer symmetric ciphers for data encryption. Because the proposed identity-based schemes are based on bilinear pairing (over elliptic curves) with large numbers, they are considerably slower compared to well-established symmetric ciphers (e.g. advanced encryption standard). Therefore, the IBSC can be modified for session key distribution with symmetric ciphers (e.g. $E_K(\cdot)$) for the actual data encryption. In the modified IBSC, the algorithms *Setup* and *Keygen* are unchanged and generate domain public parameters. The algorithm *Signcryption* is modified to encrypt a message M with a secret key K as follows:

1) Sender A randomly picks $r \xleftarrow{R} Z_q^*$ and sets

$$U = rg;$$

2) computes $h_1 = H_2(M\|K\|A\|U)$ and computes

$$V = d_A h_1 + rg_1;$$

3) computes $p_B = H_1(B\|time)$ and $h_2 = H_2(A\|B)$, and computes
$$X = h_2 U;$$

4) computes $h_3 = H_3(X\|\hat{e}(rg_1, h_2 p_B))$ and encrypts the message
$$W = K \oplus h_3;$$

5) encrypts M as $C = E_K(M)$;

6) finally outputs a 5-tuple $\langle U, V, X, W, C \rangle$.

The algorithm provides a digital signature in the same way that the original IBSC does. The consistency of the modified IBSC follows the original scheme.

12.3 Single Proxy Signing Rights Delegation

In some cases, the signing right of a specific local control center can be delegated to another local control center. A certificate is provided by the local control center itself or the PKG for signing right delegation.

12.3.1 Certificate Distribution by the Local Control Center

Let c_{ij} be the certificate of signing right delegated by L_i to L_j. A simple example of such a certificate could be $c_{ij} = AS\|j\|t_{ij}$, where t_{ij} is the expiration time of c_{ij}. A certificate can be valid for one message or for all messages before expiration of the certificate. The local control center L_i delegates c_{ij} for a message M in the following steps:

1) L_i randomly picks $y \xleftarrow{R} \mathbb{Z}_q^*$ and computes
$$U = yg;$$

2) computes $h = H_1(M\|c_{ij})$ and computes
$$V = hUp_j;$$

3) sets $W = hd_i + yg_1$.

Signing rights delegation is a 4-tuple $\sigma_c = \langle U, V, W, c_{ij} \rangle$. Once L_j receives the σ_c, it verifies σ_c if $\hat{e}(V, W) = \hat{e}(hUd_j, hp_i + U)$. The consistency is shown in the following:

$$\begin{aligned}
\hat{e}(V, W) &= \hat{e}(hUp_j, hd_i + yg_1) \\
&= \hat{e}(hUp_j, hp_i + yg)^s \\
&= \hat{e}(hUd_j, hp_i + U).
\end{aligned} \tag{12.4}$$

12.3.2 Signing Rights Delegation by the PKG

Alternatively, the PKG is able to distribute a certificate c_{ij} to L_j in the following steps:

1) The PKG randomly picks $y \in \mathbb{Z}_q^*$ and computes

$$u' = yg,$$
$$h' = H_1(m\|c_{ij}),$$
$$v' = hup_j,$$
$$w' = hd_i + hd_{AS} + yg_1;$$

2) it finally outputs a 5-tuple $\sigma_c' = \langle u', v', u', w', c_{ij}\rangle$.

The delegation σ_c' is verified by L_j if $\hat{e}(v', w') = \hat{e}(h'u'd_j, h'p_i + h'p_{AS} + u')$. The consistency is shown in the following:

$$\begin{aligned}
\hat{e}(v', w') &= \hat{e}(h'u'p_j, h'd_i + h'd_{AS} + yg_1) \\
&= \hat{e}(h'u'p_j, h'p_i + yg)^s \\
&= \hat{e}(h'u'd_j, h'p_i + h'p_{AS} + u').
\end{aligned} \quad (12.5)$$

12.3.3 Single Proxy Signature

With certificate c_{ij}, L_j is ready to sign message M on behalf of L_i in the following steps:

1) L_j randomly picks $z \xleftarrow{R} \mathbb{Z}_q^*$ and computes

$$\mu = zg,$$
$$\xi = H_1(m\|w\|c_{ij}),$$
$$\omega = w + \xi d_j + zg_1;$$

2) it finally outputs a 5-tuple $\sigma_{ij} = \{c_{ij}, u, w, \mu, \omega\}$.

The proxy signature is $\sigma_{ij} = \{c_{ij}, u, w, \mu, \omega\}$ (note that u, c_{ij}, and w are from L_i). A receiver verifies σ_{ij} by checking if the equation holds as follows:

$$\hat{e}(g, \omega) = \hat{e}(g_1, hp_i + u + \xi p_j + \mu).$$

The consistency is shown in the following,

$$\begin{aligned}
\hat{e}(g, \omega) &= \hat{e}(g, w + \xi d_j + zg_1) \\
&= \hat{e}(g, hp_i + yg + \xi p_j + \mu)^s \\
&= \hat{e}(g_1, hp_i + u + \xi p_j + \mu).
\end{aligned} \quad (12.6)$$

12.4　Group Proxy Signing Rights Delegation

The group proxy signing right of a local control center L_i is delegated by the PKG to a chosen group of local control centers (e.g. L_n for some n). Without loss of generality, assume a total number N local control centers are chosen as a group in the discussion.

12.4.1　Certificate Distribution

For each L_n in the group, the PKG generates a partial signing right certificate c_{in} and

1) randomly picks $y_n \xleftarrow{R} \mathbb{Z}_q^*$ and computes

$$u_n = y_n g,$$
$$h_n = H_1(m\|c_{in}),$$
$$v_n = h_n u p_n,$$
$$w_n = h_n d_{AS} + y_n g_1;$$

2) and finally outputs a 5-tuple $\sigma_n = \langle u_n, v_n, w_n, c_{in} \rangle$.

Once L_n receives the σ_n, it verifies the certificate by checking if $\hat{e}(v_n, w_n) = \hat{e}(hud_j, hp_{AS} + u_n)$.

12.4.2　Partial Signature

With σ_n, L_n can generate a partial signature for message M in the following steps:

1) L_n randomly picks $z_n \xi \leftarrow R\mathbb{Z}_q^*$ and computes

$$\mu_n = z_n g,$$
$$\xi_n = H_1(m\|w_n\|c_{in}),$$
$$\omega_n = w_n + \xi_n d_n + z_n g_1;$$

2) finally outputs a 5-tuple $\sigma_n' = \langle c_{in}, u_n, w_n, \mu_n, \omega_n \rangle$.

12.4.3　Group Signature

After all the proxies have generated partial signatures, one of the local control centers is chosen as the gateway (e.g. L_j) and generates a group signature in the following steps:

1) L_j computes

$$\mu_g = \sum_{n=1}^{N} \mu_n, \ \omega_g = \sum_{n=1}^{N} \omega_n, \ w_g = \sum_{n=1}^{N} w_n;$$

2) finally outputs $\sigma_g = \langle \mu_g, \omega_g, w_g, w_n \& c_{in} \ \forall n \rangle$.

A receiver verifies the group signature σ_g by checking if the equation holds as follows:

$$\hat{e}(g, \omega_g) = \hat{e}\left(g_1, \sum_{n=1}^{N}(h_n p_{AS} + \xi_n p_n) + u_g + \mu_g \right).$$

The consistency can be verified such that

$$\hat{e}(g, \omega_g) = \hat{e}\left(g, \sum_{n=1}^{N}(w_n + \xi_n d_n + z_n g_1) \right)$$

$$= \hat{e}\left(g, \sum_{n=1}^{N}(h d_{AS} + y_n g_1 + \xi_n d_n + z_n g_1) \right)$$

$$= \hat{e}\left(g_1, \sum_{n=1}^{N}(h_n p_{AS} + y_n g + \xi_n p_n + z_n g) \right)$$

$$= \hat{e}\left(g_1, \sum_{n=1}^{N}(h_n p_{AS} + \xi_n p_n) + u_g + \mu_g \right).$$

12.5 Security Analysis of the Proposed Schemes

12.5.1 Assumptions for Security Analysis

The security of the IBSC and IBS schemes is based on the following computational problems [184, 185, 188]:

- **Computational Diffie-Hellman (CDH) problem**. Given P, aP, and $bP \in \mathbb{G}_1$, for all $a, b \in \mathcal{Z}_q^*$ compute $abP \in \mathbb{G}_1$ in polynomial time.
- **Bilinear Diffie-Hellman (BDH) problem**. Given P, aP, bP, and $cP \in \mathbb{G}_1$, for all $a, b, c \in \mathcal{Z}_q^*$ compute $\hat{e}(P, P)^{abc} \in \mathbb{G}_2$ in polynomial time.

Without loss of generality, time stamp *time* is considered part of the identity *ID* in the analysis later. To make the illustration clearer, the proposed IBSC scheme is separated into an identity-based encryption scheme and identity-based signature for security analysis. Moreover, all random values are picked uniformly unless otherwise specified.

12.5.2 Identity-Based Encryption Security

12.5.2.1 Security Model

Definition 1 Semantic security for identity-based encryption (IBE) schemes [188]. If no probabilistic polynomial time adversary has a nonnegligible advantage in this game:

1) The challenger runs the setup algorithm to generate the system's parameters and sends them to the adversary.
2) The adversary $\mathcal{A}$ performs a series of queries:
   - Key extraction queries. $\mathcal{A}$ produces an identity ID and receives the private key d_{ID}.
   - Challenge. After a polynomial number of queries, $\mathcal{A}$ outputs two equal-length plaintexts M_0 and M_1 and a public key ID on which it wishes to be challenged (ID has not appeared in private key queries). The challenger picks a random bit $T \in \{0, 1\}$ and encrypts M_b according to the IBE scheme.
   - More key extraction queries. $\mathcal{A}$ issues more key extraction queries. The challenger responds as before.

Finally, the adversary $\mathcal{A}$ outputs a guess $T' \in \{0, 1\}$, $\mathcal{A}$ wins the game if $T = T'$.

12.5.2.2 Security Analysis

Lemma 12.1 Let H_1 be a random oracle from $\{0, 1\}^*$ to $\mathbb{G}_1^*$. An adversary $\mathcal{A}$ has ϵ advantage against IBE. Then there exists an adversary $\mathcal{F}$ that has advantage

$$\epsilon_{\mathcal{F}} \geq \frac{\epsilon}{e(1 + q_E)} \tag{12.7}$$

Proof: Let T be a random variable following Bernoulli distribution, that is, $T \in \{0, 1\}$ with probability δ of being 0 and probability $1 - \delta$ of being 1.

Setup. The challenger generates system parameters and sends them to $\mathcal{F}$. $\mathcal{F}$ picks a random value $p_{ID} \xleftarrow{R} \mathbb{G}_1^*$. Then $\mathcal{A}$ issues H_1 queries.

Public key queries on H_1. $\mathcal{A}$ queries oracle H_1 at ID_i, and $\mathcal{F}$ responds as follows:

- If $H_1(ID_i)$ exists in L_1 (which is a list kept by $\mathcal{F}$), then it returns that stored value.

- If $H_1(ID_i)$ does not exist in L_1, then $\mathcal{F}$ randomly chooses $T \in \{0, 1\}$ and $v_i \xleftarrow{R} \mathbb{Z}_q^*$, then sets

$$p_i = H_1(ID_i) = \begin{cases} v_i g, & T = 0 \\ v_i p_{ID}, & T = 1. \end{cases} \tag{12.8}$$

$\mathcal{F}$ stores $\langle ID_i, p_i, v_i, T \rangle$ in L_1.

Challenge: $\mathcal{A}$ outputs ID_c and two equal-length messages M_0 and M_1. $\mathcal{F}$ gives the challenger M_0, M_1; the challenger then randomly picks $u \in \{0, 1\}$ and encrypts $M_u \to C = \langle X, W \rangle$. $\mathcal{F}$ runs the response to H_1 queries to find p_c such that $H_1(ID_c) = p_c$. Then $\mathcal{F}$ finds $\langle ID_c, p_c, v_i, T \rangle$.

- If $T = 0$, then $\mathcal{F}$ aborts.
- If $T = 1$, then we have $p_c = v_i p_{ID}$. Set $C' = \langle v_i^{-1} X, W \rangle$. Since

$$\begin{aligned} \hat{e}(v_i^{-1} X, d_c) &= \hat{e}(X, v_i^{-1} s p_c) \\ &= \hat{e}(X, v_i^{-1} s v_i p_{ID}) \\ &= \hat{e}(X, d_{ID}); \end{aligned}$$

therefore, C' is IBE of M_u under ID_c, and decryption of C' using d_c is the same as decryption of C using d_{ID}.

Guess: $\mathcal{F}$ outputs a guess u'.

If $\mathcal{F}$ does not abort during the process, then $|Pr[u = u'] - 1/2| \geq \epsilon$, where the probability is over the random bits used by $\mathcal{A}$, $\mathcal{F}$ and the challenger. Let

- E_1 be the event that $\mathcal{F}$ aborts in private key queries,
- E_2 be the event that $\mathcal{F}$ aborts in challenge stage.

Then we have the probability of not aborting is

$$Pr(\neg E_1 \wedge \neg E_2) = Pr(\neg E_1) Pr(\neg E_2) = \delta^{q_E} (1 - \delta). \tag{12.9}$$

The maximum probability is achieved at $\delta_{opt} = \frac{q_E}{q_E + 1}$, which implies that

$$Pr(\neg E_1 \wedge \neg E_2) = \delta^{q_E} (1 - \delta) \geq \frac{1}{e(1 + q_E)}. \tag{12.10}$$

Thus, we can conclude that $\mathcal{F}$ has an advantage $\epsilon_{\mathcal{F}} \geq \frac{\epsilon}{e(1 + q_E)}$. □

Lemma 12.2 Let H_3 be a random oracle from $\mathbb{Z}_q^*$ to $\{0, 1\}^n$. Then there is an algorithm $\mathcal{F}$ that solves the BDH problem with advantage

$$\epsilon_B \geq \frac{2\epsilon}{q_{H_3}}. \tag{12.11}$$

Proof: Given $(g, P_1, P_2, P_3) = (g, ag, bg, cg)$, $\mathcal{F}$ sets $g_1 = P_1$ and $p_{ID} = P_2$. $\mathcal{A}$ then issues H_3 queries for h_i.

H_3 *queries.* Suppose that $\mathcal{F}$ keeps a list L_2 for $\langle h_i, H_3(h_i) \rangle$. If $H_3(h_i)$ exists in L_2, return that value. Otherwise, $\mathcal{F}$ randomly picks $v \xleftarrow{R} \{0,1\}^n$ and sets $H_i(h_i) = v$.

Challenge. $\mathcal{A}$ outputs M_0 and M_1. $\mathcal{F}$ randomly picks $Y \xleftarrow{R} \{0,1\}^n$ and defines $C = \langle P_3, Y \rangle$. $\mathcal{F}$ gives C to $\mathcal{A}$. Note that by definition, the decryption is

$$Y \oplus H_3(Y \| \hat{e}(Y, d_{ID})) = Y \oplus H_3(Y \| D),$$

where $D = \hat{e}(Y, d_{ID})$.

Guess: $\mathcal{A}$ outputs $u' \xleftarrow{R} \{0,1\}$. $\mathcal{F}$ randomly picks $\langle h_j, H_3(h_j) \rangle \xleftarrow{R} L_2$ and outputs h_j. Note that with Y and h_j, D can be computed since $Y \| h_j = D$, where D is the solution to the BDH problem.

Let E_H be the event that $\mathcal{A}$ issues H_3 queries for $H_3(Y \| D)$, then from [188] we know that $Pr[E_H] \geq \epsilon$ and thus $\epsilon_B \geq \frac{2\epsilon}{q_{H_3}}$. $\qquad\square$

Theorem 12.1 Suppose H_1 and H_3 are random oracles, $\mathcal{A}$ has advantage ϵ against IBE within running time t. $\mathcal{A}$ also makes q_E private key extraction queries and q_{H_3} H_3 queries. Then there exists polynomial algorithm $\mathcal{F}$ that solves the BDH problem with advantage

$$\epsilon' \geq \frac{2\epsilon}{e(1 + q_E)q_{H_3}}, \tag{12.12}$$

within a time $t' < t + (q_{H_1} + q_E + q_{H_3})t_m$, where t_m is the time to compute a scalar multiplication in $\mathbb{G}_1^*$.

Proof: Theorem 12.1 follows Lemma 12.1 and Lemma 12.2 directly. $\square$

12.5.3 Identity-Based Signature Security

12.5.3.1 Security Models
Definition 2 Strongly existentially unforgeable identity-based signature scheme under chosen-message attacks [184]. If no probabilistic polynomial time adversary has a nonnegligible advantage in this game:

1) The challenger runs the setup algorithm to generate the system's parameters and sends them to the adversary.

2) The adversary $\mathcal{A}$ performs a series of queries:
   - Key extraction queries. $\mathcal{A}$ produces an identity ID_i and receives the private key d_i.
   - Signature queries $\mathcal{A}$ produces a message M and an identity ID_i and receives a signature on M that was generated by the signature oracle using the private key corresponding to the identity ID_i (i.e. d_i).
   - After a polynomial number of queries, $\mathcal{A}$ produces a tuple (ID, M, σ) made of an identity ID, whose corresponding private key was never asked during stage 2, and a message signature pair (M, σ) such that σ was not returned by the signature oracle on the input (M, ID) during stage 2 for the identity ID.

$\mathcal{A}$ wins the game if the forged signature can be verified when the verification algorithms run on the tuple (ID, M, σ). The forger's advantage is defined to be its probability of producing a forgery taken over the number of coin-flipping of the challenger and $\mathcal{A}$.

12.5.3.2 Security Analysis

Theorem 12.2 Let H_1 and H_2 be random oracles, $\mathcal{A}$ has advantage ϵ against IBS in running time t. $\mathcal{A}$ also makes q_E private key extraction queries, q_{H_2} H_2 queries, and q_S signature queries. Then there is an algorithm $\mathcal{F}$ that solves the CDH problem with advantage

$$\epsilon_C \geq \frac{\epsilon - q_S(q_{H_2} + q_S)/q}{e(q_E + 1)} \tag{12.13}$$

within running time $t' < t + (q_{H_1} + q_E + q_{H_2} + 2q_S)t_m + (q_S + 1)t_{mm}$, where t_m is the running time for a scalar multiplication in $\mathbb{G}_1^*$, and t_{mm} is the running time for a multiexponentiation in $\mathbb{G}_1^*$.

Proof: Let $P_1 = ag$ and $P_2 = bg$ be the input of the CDH problem. Given (P_1, P_2). Fist, $\mathcal{F}$ initializes $g_1 = P_1$ as the domain secret. From the perspective of the adversary, the distribution of g_1 secret is identical to the real one (i.e. $g_1 = sg$). Let T be a random variable following Bernoulli distribution, i.e. $T \in \{0, 1\}$ with probability δ of being 0, and probability $1 - \delta$ of being 1. Then, $\mathcal{F}$ issues a series of queries as stated in the following:

1) H_1 *queries.* Suppose the adversary issues a query for an identity ID_i. $\mathcal{F}$ first picks a random number $u_i \xleftarrow{R} \mathbb{Z}_q^*$ and decides the public key

based on the outcome of T, such that

$$p_i = \begin{cases} u_i P_2, & T = 1, \\ u_i g, & T = 0. \end{cases} \tag{12.14}$$

$\mathcal{F}$ then keeps (ID_i, u_i, T) in list L_1.

2) *Private key queries.* For ID_i, $\mathcal{F}$ recovers u_i and T from L_1, such that

$$(ID_i, u_i, T) \leftarrow L_1.$$

And the private key of ID_i is determined as

$$d_i = \begin{cases} Abort, & T = 1, \\ u_i P_1, & T = 0. \end{cases} \tag{12.15}$$

$T = 1$ indicates that there is no answer to the query. When $T = 0$, note that $d_i = ap_i$ follows the distribution of the real secret key.

3) H_2 *queries.* Assuming that $\mathcal{F}$ keeps a list L_2 that stores any previously defined $H_2(h_i)$. Given (ID_i, M, U_i), $\mathcal{F}$ first checks whether $H_2(h_i)$ has been defined (e.g. H_i). If so, the defined value will be returned. Otherwise, $\mathcal{F}$ randomly picks $v_i \xleftarrow{R} \mathbb{Z}_q^*$, and determines $H_2(h_i)$ as

$$H_2(h_i) = \begin{cases} v_i, & \text{not defined in } L_2, \\ H_i, & \text{already defined in } L_2. \end{cases} \tag{12.16}$$

4) *Signature queries.* $\mathcal{A}$ randomly chooses $\mu_i \xleftarrow{R} \mathbb{Z}_q^*$ and $v_i \xleftarrow{R} \mathbb{Z}_q^*$. Then set $U_i = \mu_i g$ and $V_i = v_i g_1$. Define $H_2(h_i) = (u_i P_2)^{-1}(v_i g - U_i) \in \mathbb{Z}_q^*$. The pair $\sigma_i = \langle U_i, V_i \rangle$ appears as a valid signature.

$$\sigma_i = \begin{cases} \langle U_i, V_i \rangle, & H_2(h_i) \text{ not defined in } L_2, \\ Abort, & H_2(h_i) \text{ defined in } L_2. \end{cases} \tag{12.17}$$

5) *Signature forgery.* Given a message M and an identity ID, $\mathcal{A}$ forges a signature $\langle U, V \rangle$. $\mathcal{F}$ recovers

$$(ID, u, T) \leftarrow L_1.$$

If $T = 0$, then abort. Otherwise (i.e. $T = 1$), the list L_2 must contain an entry (ID, M, U, v) with overwhelming probability. Since $H_2(h_{ID})$ has been defined as v, if $\mathcal{A}$ succeeds in the game, then $\mathcal{F}$ knows that

$$\hat{e}(g, V) = \hat{e}(g_1, p_{ID} h_{ID} + U).$$

With $h_{ID} = v, p_{ID} = uP_2$, where u and v are known, then $\mathcal{F}$ also finds that

$$\hat{e}(g, V) = \hat{e}(g_1, p_{ID}h_1 + U)$$
$$= \hat{e}(g_1, p_{ID}h_{ID})\hat{e}(g_1, U)$$
$$= \hat{e}(P_1, uP_2v)\hat{e}(P_1, U)$$
$$\Rightarrow \hat{e}(g, V - vP_1) = \hat{e}(P_1, vuP_2).$$

And $(vu)^{-1}(V - vP_1)$ is the solution to the CDH instance (P_1, P_2).

From Lemma 12.1 we know that the probability of $\mathcal{F}$ not aborting in the process of key extraction query is at least $1/e(q_E + 1)$. Moreover, the probability that $\mathcal{F}$ aborts in the process of signature queries is at most $q_S(q_{H_2} + q_S)/q$ due to conflict on H_2, where q is the size of $\mathbb{G}_1$. Overall, $\mathcal{F}$ has an advantage at least $\epsilon' \geq \frac{\epsilon - q_S(q_{H_2} + q_S)/q}{e(q_E + 1)}$. That completes the proof. □

12.6 Performance Analysis of the Proposed Schemes

In this section, we evaluate the performance of the proposed schemes.

12.6.1 Computational Complexity of the Proposed Schemes

Performance of the proposed schemes is based on the number of operations and the efficiency of each type of operations. Table 12.1 lists the number of operations of each algorithm in the proposed security scheme. Among them, *mul* indicates standard multiplication in $\mathbb{G}_1$. Since addition in $\mathbb{G}_1$ and XOR are simple and efficient operations, they are not listed in the table.

Hash functions can be computed efficiently in general. In practice, H_2 and H_3 are easy to find. However, it is hard to build $H_1 : \{0, 1\} \rightarrow \mathbb{G}_1^*$. In the analysis, we relax H_1 into two steps as follows:

1) $H_1 : \{0, 1\}^* \rightarrow I \subseteq \{0, 1\}^*$;
2) $H_1' : I \rightarrow \mathbb{G}_1^*$.

In step 1, I is a finite set, and H_1' is an encoding function which is computable. Note that after the relaxation, the public key for a given *ID* and *time* is calculated as $p_{ID} = H_1'(H_1(ID\|time))$. In the proposed IBSC

Table 12.1 Computational complexity.

	# of $\hat{e}$	# of *mul*	# of H_1	# of H_2	# of H_3
Signcrypt	1	5	1	2	1
Decrypt	1	0	0	0	1
Sign	0	3	1	0	0
Verify	3	1	1	1	0

scheme, public keys can be computed at the beginning of each session and cached for the entire session. Therefore, the relaxation of H_1 does not introduce more computational cost in reality. Because of that, performance of the IBSC and IBS schemes will be considered efficient if bilinear pairing $\hat{e}$ and multiplication in $\mathbb{G}_1$ can be computed efficiently. Since the Weil pairing can be performed efficiently using Miller's algorithm [191], the bilinear map $\hat{e}$ can be performed efficiently as well.

12.6.2 Choosing Bilinear Paring Functions

To analyze the performance of the IBSC scheme, we apply two bilinear pairing functions, *modified Weil pairing* and *Tate pairing*, over supersingular elliptic curve $E : \{y^2 = x^3 + 1 | x, y \in \mathbb{F}_p\}$.

We first construct $\mathbb{G}_1$. Let p be a prime number such that $p \equiv 2 \bmod 3$ and $p = aq - 1$ for some prime q and positive integer a. Then $\mathbb{G}_1$ is the subgroup of order q of $\mathbb{F}_{p^2}^*$. The CDH problem is hard in the group $\mathbb{G}_1$ [188, 192]. However, it is worth mentioning that the decisional Diffie-Hellman (DDH) problem is an easy one for bilinear map $\hat{e}$. This is because with given $P, aP, bP, cP \in \mathbb{G}_1, \forall a, b, c \in \mathbb{Z}_q^*$, we can easily check if $c \equiv ab \bmod q$ by comparing $\hat{e}(aP, bP)$ with $\hat{e}(P, cP)$.

> **Decisional Diffie-Hellman problem (DDHP).** Given $P, aP, bP, cP \in \mathcal{G}_1$, for all $a, b, c \in \mathcal{Z}_q^*$, decide whether $c \equiv ab \bmod q$.

The Weil pairing e has the properties of bilinearity and computability; however, it does not have nondegeneracy. Therefore, we adopt a modified Weil pairing $\hat{e} : \mathbb{G}_1 \times \mathbb{G}_1 \to \mathbb{G}_2$ such that $\hat{e}(P, Q) = e(P, \phi(Q))$, where ϕ is an automorphism on the group of points of supersingular elliptic curve $E : \{y^2 = x^3 + 1 | x, y \in \mathbb{F}_p\}$, that is, $\phi(x, y) = (\xi x, y)$, where ζ is a primitive cube root of unity in $\mathbb{F}_p$. Thus, $y^2 = (\zeta x)^3 + 1 = \zeta^3 x^3 + 1 = x^3 + 1 \Rightarrow \phi(P_1) + \phi(P_2) = \phi(P_1 + P_2), \forall P_1, P_2 \in \mathbb{G}_1$. The bilinear map $\hat{e}$ is calculated as a Weil pairing with an additional standard multiplication on the curve E. According to [188], $\hat{e}$ is believed to satisfy the BDH problem. However, computing the discrete logarithm in $\mathbb{F}_p^*$ is sufficient for computing the discrete logarithm in $\mathbb{G}_1$. Therefore, in order to make it sufficiently hard in practice, q needs to be at least 512 bits long.

12.6.3 Numerical Results

We evaluate the proposed identity-based schemes with modified Weil pairing $\hat{e}$ using Mathematica 10.0 on a computer equipped with an Intel Core i5-2400 @ 3.1 GHz and 12 GB RAM. We first show the computational cost of each operation. Since H_1, H_2, and H_3 do not have much difference in computational time and the added encoding function H_1' is more efficient than H_1, the hash functions are excluded from the performance analysis.

First, we evaluate the computational time for the bilinear pairing $\hat{e}$. Two sets of evaluation are given, that is, for $q = 256$ bits and $q = 385$ bits. Each evaluation is the average value from 10,000 calculations. With $q = 256$ bits, one $\hat{e}$ takes about 7.44 ms. With $q = 385$ bits, one $\hat{e}$ takes about 13.25 ms. We then evaluate the computational time of standard multiplication over $\mathbb{G}_1$ (i.e. $k_p P \in \mathbb{G}_1$). The computational time of $k_p P \in \mathbb{G}_1$ mainly depends on the size of k_p (assuming $q = 256$ bits). The computational time of each evaluation is averaged from 10,000 calculations. With $k_p = 128/256/512$ bits, a standard multiplication operation takes about $3.25/6.43/12.29$ ms. Note that in the proposed IBSC, k_p is the output of some hash functions; therefore, k_p usually is 256 bits or 512 bits, where the computation is efficient. The evaluation results are summarized in Table 12.2.

Based on the evaluations we have for each operation, we then show the total operational time for each algorithm. In practice, public keys are computed once and cached for the entire session.

Table 12.2 Computational time for each operation.

Bilinear pairing $\hat{e}$		Standard multiplication	
$q = 256$ bits	$q = 385$ bits	$k_p = 256$ bits	$k_p = 512$ bits
7.44 ms	13.25 ms	6.43 ms	12.29 ms

Table 12.3 Computational time of each algorithm.

	$q = 256\,b$ $k_p = 256\,b$	$q = 256\,b$ $k_p = 512\,b$	$q = 385\,b$ $q = 256\,b$	$q = 385\,b$ $q = 512\,b$
Signcrypt	39.59 ms	68.89 ms	45.4 ms	74.7 ms
Decrypt	7.44 ms	7.44 ms	13.25 ms	13.25 ms
Sign	19.29 ms	36.87 ms	19.29 ms	36.87 ms
Verify	28.75 ms	34.61 ms	46.18 ms	52.04 ms

The computational time of each algorithm is listed in Table 12.3 (where "b" stands for "bits"). It is shown that the proposed IBSC performs efficiently for delay-tolerant and even near real-time data transmission, for example, metering data transmission. However, for real-time monitoring data, such as PMU data, the identity-based schemes alone may not be a good solution. Without sufficient computational resources, faster security protocols and schemes are recommended, for instance, traditional symmetric ciphers. The proposed IBSC can be applied for initial authentication and key distribution of the chosen symmetric ciphers.

12.7 Conclusion

In this chapter, we proposed an ID-based signcryption security scheme for smart grid communications over public networks. The proposed IBSC scheme performs simultaneously the functions of encryption and digital signature. Therefore, confidentiality, non-repudiation, and data integrity are provided in a single calculation. The proposed IBSC scheme was also reduced to a ID-based digital signature scheme if confidentiality is not required for some messages. To further enhance the performance, symmetric ciphers were introduced to the IBSC. In addition, delegation of signing rights from one local control center to another (or a few) local control center was achieved by the proposed identity-based schemes. The security of the proposed IBSC was studied. The numerical results showed that the proposed IBSC scheme is able to perform efficiently with security guarantee in the cyber-physical system of the smart grid.

13

Open Issues and Possible Future Research Directions

There are still several challenges and issues to be addressed in the smart grid communications infrastructure. In this chapter, we will present some of the open issues and point out possible directions for future research.

13.1 Efficient and Secure Cloud Services and Big Data Analytics

In the proposed information communications technology (ICT) framework, cloud services and big data analytics will be critical components of the smart grid communication infrastructure. Considering the amount of data and the scale of the grid, ways to achieve efficient and cost-effective cloud services have yet to be studied. The exact techniques of big data analytics also need to be developed specifically for metering data as well as useful data from external sources. With those techniques, accurate price and energy forecasts can be made to enhance demand response and the operational efficiency of the grid as well. Besides efficiency, security is an equally important issue to be considered when developing cloud services and techniques for big data analytics so that the privacy of customers can be protected.

13.2 Quality-of-Service Framework

The quality-of-service (QoS) in the smart grid can be defined by the accuracy and effectiveness with which different pieces of information (such as the state of equipment, load information, and power pricing)

Smart Grid Communication Infrastructures: Big Data, Cloud Computing, and Security,
First Edition. Feng Ye, Yi Qian, and Rose Qingyang Hu.
© 2018 John Wiley & Sons Ltd. Published 2018 by John Wiley & Sons Ltd.

Table 13.1 Latency requirements in smart grid communications.

Maximum latency	Communication type
$\leq 4\ ms$	Protective relaying
Sub-seconds	Wide area situational awareness monitoring
Seconds	Substation and feeder supervisory control and data acquisition (SCADA)
Minutes	Monitoring noncritical equipment and marketing pricing information
Hours	Meter reading and longer-term pricing information
Days/Weeks/Months	Collecting long-term usage data

are delivered in a timely manner to the respective parties. The QoS framework can be developed by identifying the specific QoS requirements and priorities for specific communication networks in the smart grid. For example, consider the latency requirements listed in Table 13.1.

13.3 Optimal Network Design

The smart grid will be deployed on a large scale with an increasing number of customers included. More data will be gathered and transmitted more frequently. For example, current metering data is gathered once in 15 minutes; however, a higher gathering frequency is needed for real-time demand response. Therefore, the communication networks in smart grid will need to be enhanced to achieve better network performance that will meet the increasing demand for transmission of data. Technologies in smart grid communications are not separated from other communications. Many cutting-edge technologies can be developed and deployed in the smart grid network design as well. For example, relaying, cooperative communications and small cells are being developed in the next generation of mobile communications networks can also be deployed for networks in the smart grid, especially home-area networks, neighborhood-area networks, and wide-area monitoring systems. Those technologies will result in more efficient networks with larger coverage that are an important part of smart grid communications.

13.4   Better Involvement of Green Energy

Green energy or renewable resources have attracted a lot of atten-
tion recently. Several photovoltaic (PV) farms and wind farms have
been deployed all over the world. However, the deployment, operational, and maintenance costs are still high compared with fossil-fuel
power stations. In the smart grid, a higher proportion of green energy
is vital for lower greenhouse gas emissions. Research focusing on better involvement of green energy needs to be addressed. Some topics that need to be studied in the near future include management of
renewable resources, deployment and management of energy storage
units, and the involvement of microgrids.

13.5   Need for Secure Communication Network Infrastructure

If the smart grid is attacked, hackers can penetrate the network
and alter critical system parameters, which could destabilize the
grid unpredictably and cause a nationwide crisis. Security must
be designed for various smart grid systems to meet their different
requirements. Research directions include intrusion detection system
(IDS) and intrusion prevention system (IPS) for smart communications infrastructure, for example, AMI and WAMS. IDS and IPS
are particularly important in the area of big data analytics. Key
management (e.g. public-key infrastructure) should be studied as the
smart grid evolves to secure not only new, and modern equipment
but also legacy devices.

13.6   Electrical Vehicles

Electrical vehicles (EVs) and plug-in hybrid electric vehicles (PHEVs)
will consume a large portion of electricity in the near future. Several
issues are raised because of EVs and PHEVs. For instance, what is the
best way to manage the charging and discharging of those EVs and
PHEVs so that demand response can still remain optimal for smoothing the power load? More power plants may be needed to compensate
for the extra electricity requirements from EVs/PHEVs. However, that

may not necessarily be the case, because EVs/PHEVs can also be power suppliers for the power grid. Together, EVs/PHEVs store a considerable amount of electricity. Unfortunately, utilizing such storage efficiently with proper controls still remains an open issue, because EVs/PHEVs are not regular appliances that remain in one place. Moreover, data collection and exchange will most likely go through wireless networks, and EVs/PHEVs tend to stay still for a relatively long period of time when charging and discharging. That raises the issue of security. How can the privacy of users of EVs/PHEVs be protected when exchanging useful information? Those issues need to be addressed before the roads are crowded with EVs/PHEVs in the near future.

Reference

1 Niyato, D., Wang, P., Han, Z., and Hossain, E. (2011) Impact of packet loss on power demand estimation and power supply cost in smart grid, in *Wireless Communications and Networking Conference (WCNC), 2011 IEEE*, pp. 2024–2029, doi:10.1109/WCNC.2011.5779440.

2 Amin, S.M. (2012) Smart grid security, privacy, and resilient architectures: Opportunities and challenges, in *Power and Energy Society General Meeting, 2012 IEEE*, IEEE, pp. 1–2.

3 Niyato, D., Dong, Q., Wang, P., and Hossain, E. (2013) Optimizations of power consumption and supply in the smart grid: Analysis of the impact of data communication reliability. *Smart Grid, IEEE Transactions on*, 4 (1), 21–35, doi:10.1109/TSG.2012.2224677.

4 Yan, Y., Qian, Y., Sharif, H., and Tipper, D. (2012) A survey on cyber security for smart grid communications. *IEEE Communications Surveys Tutorials*, 14 (4), 998–1010, doi:10.1109/SURV.2012.010912.00035.

5 Fang, X., Misra, S., Xue, G., and Yang, D. (2012) Smart grid–2014; the new and improved power grid: A survey. *Communications Surveys Tutorials, IEEE*, 14 (4), 944–980, doi:10.1109/SURV.2011.101911.00087.

6 Zhou, L., Rodrigues, J., and Oliveira, L. (2012) Qoe-driven power scheduling in smart grid: architecture, strategy, and methodology. *Communications Magazine, IEEE*, 50 (5), 136–141, doi:10.1109/MCOM.2012.6194394.

Smart Grid Communication Infrastructures: Big Data, Cloud Computing, and Security,
First Edition. Feng Ye, Yi Qian, and Rose Qingyang Hu.
© 2018 John Wiley & Sons Ltd. Published 2018 by John Wiley & Sons Ltd.

7 Yan, Y., Qian, Y., Sharif, H., and Tipper, D. (2013) A survey on smart grid communication infrastructures: Motivations, requirements and challenges. *IEEE Communications Surveys Tutorials*, **15** (1), 5–20, doi:10.1109/SURV.2012.021312.00034.

8 Zhou, L. and Rodrigues, J. (2013) Service-oriented middleware for smart grid: Principle, infrastructure, and application. *Communications Magazine, IEEE*, **51** (1), 84–89, doi:10.1109/MCOM.2013.6400443.

9 Erol-Kantarci, M. and Mouftah, H. (2015) Energy-efficient information and communication infrastructures in the smart grid: A survey on interactions and open issues. *Communications Surveys Tutorials, IEEE*, **17** (1), 179–197, doi:10.1109/COMST.2014.2341600.

10 Ye, F., Qian, Y., and Hu, R. (2015) Energy efficient self-sustaining wireless neighborhood area network design for smart grid. *Smart Grid, IEEE Transactions on*, **6** (1), 220–229, doi:10.1109/TSG.2014.2344659.

11 Smart grid., https://www.smartgrid.gov. [Online; accessed 8-July-2017].

12 Fossil Fuel Power Station, http://en.wikipedia.org/wiki/Fossil-fuel_power_station. [Online; accessed 8-June-2015].

13 Ma, Z., Callaway, D., and Hiskens, I. (2010) Decentralized charging control for large populations of plug-in electric vehicles, in *Decision and Control (CDC), 2010 49th IEEE Conference on*, pp. 206–212, doi:10.1109/CDC.2010.5717547.

14 Fan, Z. (2012) A distributed demand response algorithm and its application to phev charging in smart grids. *Smart Grid, IEEE Transactions on*, **3** (3), 1280–1290, doi:10.1109/TSG.2012.2185075.

15 Salinas, S., Li, M., and Li, P. (2013) Privacy-preserving energy theft detection in smart grids: A p2p computing approach. *Selected Areas in Communications, IEEE Journal on*, **31** (9), 257–267, doi:10.1109/JSAC.2013.SUP.0513023.

16 Jiang, R., Lu, R., Wang, Y., Luo, J., Shen, C., and Shen, X.S. (2014) Energy-theft detection issues for advanced metering infrastructure in smart grid. *Tsinghua Science and Technology*, **19** (2), 105–120, doi:10.1109/TST.2014.6787363.

17 Gridwise Alliance, The future of the grid, https://www.smartgrid.gov/files/Future_of_the_Grid_web_final_v2.pdf. [Online; accessed 9-July-2017].

18 Erol-Kantarci, M. and Mouftah, H. (2011) Wireless sensor networks for cost-efficient residential energy management in the smart grid. *Smart Grid, IEEE Transactions on*, **2** (2), 314–325, doi:10.1109/TSG.2011.2114678.

19 McBee, K. and Simoes, M. (2012) Utilizing a smart grid monitoring system to improve voltage quality of customers. *Smart Grid, IEEE Transactions on*, **3** (2), 738–743, doi:10.1109/TSG.2012.2185857.

20 Qiu, M., Su, H., Chen, M., Ming, Z., and Yang, L. (2012) Balance of security strength and energy for a pmu monitoring system in smart grid. *Communications Magazine, IEEE*, **50** (5), 142–149, doi:10.1109/MCOM.2012.6194395.

21 Baki, A. (2014) Continuous monitoring of smart grid devices through multi protocol label switching. *Smart Grid, IEEE Transactions on*, **5** (3), 1210–1215, doi:10.1109/TSG.2014.2301723.

22 (2014), NIST Framework and Roadmap for Smart Grid Interoperability Standards, Release 3.0, http://www.nist.gov/public_affairs/releases/upload/smartgrid_interoperability_final.pdf. [Online; accessed 9-July-2017].

23 Güzelgöz, S., Arslan, H., Islam, A., and Domijan, A. (2011) A review of wireless and plc propagation channel characteristics for smart grid environments. *JECE*, **2011**, 15:15–15:15, doi:10.1155/2011/154040. URL http://dx.doi.org/10.1155/2011/154040.

24 Galli, S., Scaglione, A., and Wang, Z. (2011) For the grid and through the grid: The role of power line communications in the smart grid. *Proceedings of the IEEE*, **99** (6), 998–1027, doi:10.1109/JPROC.2011.2109670.

25 Gomez-Cuba, F., Asorey-Cacheda, R., and Gonzalez-Castano, F. (2013) Smart grid last-mile communications model and its application to the study of leased broadband wired-access. *Smart Grid, IEEE Transactions on*, **4** (1), 5–12, doi:10.1109/TSG.2012.2223765.

26 Pagani, G. and Aiello, M. (2013) Modeling the last mile of the smart grid, in *Innovative Smart Grid Technologies (ISGT), 2013 IEEE PES*, pp. 1–6, doi:10.1109/ISGT.2013.6497816.

27 Ma, J., Deng, J., Song, L., and Han, Z. (2014) Incentive mechanism for demand side management in smart grid using auction. *Smart Grid, IEEE Transactions on*, **5** (3), 1379–1388, doi:10.1109/TSG.2014.2302915.

28 Soliman, H. and Leon-Garcia, A. (2014) Game-theoretic demand-side management with storage devices for the future smart grid. *Smart Grid, IEEE Transactions on,* **5** (3), 1475–1485, doi:10.1109/TSG.2014.2302245.

29 Fadlullah, Z., Quan, D.M., Kato, N., and Stojmenovic, I. (2014) Gtes: An optimized game-theoretic demand-side management scheme for smart grid. *Systems Journal, IEEE,* **8** (2), 588–597, doi:10.1109/JSYST.2013.2260934.

30 Li, S., Zhang, D., Roget, A., and O'Neill, Z. (2014) Integrating home energy simulation and dynamic electricity price for demand response study. *Smart Grid, IEEE Transactions on,* **5** (2), 779–788, doi:10.1109/TSG.2013.2279110.

31 Fadlullah, Z., Fouda, M., Kato, N., Takeuchi, A., Iwasaki, N., and Nozaki, Y. (2011) Toward intelligent machine-to-machine communications in smart grid. *Communications Magazine, IEEE,* **49** (4), 60–65, doi:10.1109/MCOM.2011.5741147.

32 Niyato, D., Xiao, L., and Wang, P. (2011) Machine-to-machine communications for home energy management system in smart grid. *Communications Magazine, IEEE,* **49** (4), 53–59.

33 Liu, E., Chan, M., Huang, C., Wang, N., and Lu, C. (2010) Electricity grid operation and planning related benefits of advanced metering infrastructure, in *Critical Infrastructure (CRIS), 2010 5th International Conference on,* pp. 1–5, doi:10.1109/CRIS.2010.5617583.

34 Zhou, J., Hu, R., and Qian, Y. (2012) Scalable distributed communication architectures to support advanced metering infrastructure in smart grid. *Parallel and Distributed Systems, IEEE Transactions on,* **23** (9), 1632–1642, doi:10.1109/TPDS.2012.53.

35 Gharavi, H. and Xu, C. (2012) Traffic scheduling technique for smart grid advanced metering applications. *Communications, IEEE Transactions on,* **60** (6), 1646–1658, doi:10.1109/TCOMM.2012.12.100620.

36 Ye, F., Qian, Y., Hu, R.Q., and Das, S.K. (2015) Reliable energy-efficient uplink transmission for neighborhood area networks in smart grid. *IEEE Transactions on Smart Grid,* **6** (5), 2179–2188.

37 Bruce, A. (1998) Reliability analysis of electric utility scada systems. *Power Systems, IEEE Transactions on,* **13** (3), 844–849, doi:10.1109/59.708711.

38 Patel, M., Aivaliotis, S., Ellen, E. *et al.* (2010) Real-time application of synchrophasors for improving reliability. *NERC Report,* Oct.

39 Xie, Z., Manimaran, G., Vittal, V., Phadke, A., and Centeno, V. (2002) An information architecture for future power systems and its reliability analysis. *Power Systems, IEEE Transactions on,* **17** (3), 857–863, doi:10.1109/TPWRS.2002.800971.

40 Wang, Y., Li, W., and Lu, J. (2010) Reliability analysis of wide-area measurement system. *Power Delivery, IEEE Transactions on,* **25** (3), 1483–1491, doi:10.1109/TPWRD.2010.2041797.

41 Shahraeini, M., Javidi, M., and Ghazizadeh, M. (2011) Comparison between communication infrastructures of centralized and decentralized wide area measurement systems. *Smart Grid, IEEE Transactions on,* **2** (1), 206–211, doi:10.1109/TSG.2010.2091431.

42 Wang, Y., Wang, C., Li, W., Li, J., and Lin, F. (2014) Reliability-based incremental pmu placement. *Power Systems, IEEE Transactions on,* **29** (6), 2744–2752, doi:10.1109/TPWRS.2014.2310182.

43 Ray, P., Harnoor, R., and Hentea, M. (2010) Smart power grid security: A unified risk management approach, in *Security Technology (ICCST), 2010 IEEE International Carnahan Conference on,* pp. 276–285, doi:10.1109/CCST.2010.5678681.

44 Komninos, N., Philippou, E., and Pitsillides, A. (2014) Survey in smart grid and smart home security: Issues, challenges and countermeasures. *Communications Surveys Tutorials, IEEE,* **16** (4), 1933–1954, doi:10.1109/COMST.2014.2320093.

45 Maharjan, S., Zhu, Q., Zhang, Y., Gjessing, S., and Basar, T. (2013) Dependable demand response management in the smart grid: A stackelberg game approach. *Smart Grid, IEEE Transactions on,* **4** (1), 120–132, doi:10.1109/TSG.2012.2223766.

46 Ye, F., Qian, Y., and Hu, R. (2015 (early access)) A real-time information based demand-side management system in smart grid. *Parallel and Distributed Systems, IEEE Transactions on,* **PP** (99), 1–1, doi:10.1109/TPDS.2015.2403833.

47 Lasseter, R.H. (2002) Microgrids, in *Power Engineering Society Winter Meeting, 2002. IEEE,* vol. 1, IEEE, vol. **1**, pp. 305–308.

48 Erol-Kantarci, M., Kantarci, B., and Mouftah, H. (2011) Reliable overlay topology design for the smart microgrid network. *Network, IEEE,* **25** (5), 38–43, doi:10.1109/MNET.2011.6033034.

49 Liang, H., Choi, B.J., Abdrabou, A., Zhuang, W., and Shen, X. (2012) Decentralized economic dispatch in microgrids via heterogeneous wireless networks. *Selected Areas in Communications, IEEE Journal on*, **30** (6), 1061–1074, doi:10.1109/JSAC.2012.120705.

50 Kanabar, M., Adamiak, M., and Rodrigues, J. (2013) Optimizing wide area measurement system architectures with advancements in phasor data concentrators (pdcs), in *Power and Energy Society General Meeting (PES), 2013 IEEE*, pp. 1–5, doi:10.1109/PESMG.2013.6672987.

51 Chuang, C.L., Wang, Y.C., Lee, C.H., Liu, M.Y., Hsiao, Y.T., and Jiang, J.A. (2010) An adaptive routing algorithm over packet switching networks for operation monitoring of power transmission systems. *Power Delivery, IEEE Transactions on*, **25** (2), 882–890, doi:10.1109/TPWRD.2008.2008494.

52 Huang, J., Wang, H., Qian, Y., and Wang, C. (2013) Priority-based traffic scheduling and utility optimization for cognitive radio communication infrastructure-based smart grid. *Smart Grid, IEEE Transactions on*, **4** (1), 78–86, doi:10.1109/TSG.2012.2227282.

53 Ma, R., Meng, W., Chen, H.H., and Huang, Y.R. (2012) Coexistence of smart utility networks and wlan/zigbee in smart grid, in *Smart Grid Communications (SmartGridComm), 2012 IEEE Third International Conference on*, pp. 211–216, doi:10.1109/SmartGridComm.2012.6485985.

54 Chang, K.H. and Mason, B. (2012) The ieee 802.15.4g standard for smart metering utility networks, in *Smart Grid Communications (SmartGridComm), 2012 IEEE Third International Conference on*, pp. 476–480, doi:10.1109/SmartGridComm.2012.6486030.

55 Li, Z. and Liang, Q. (2013) Performance analysis of multiuser selection scheme in dynamic home area networks for smart grid communications. *Smart Grid, IEEE Transactions on*, **4** (1), 13–20, doi:10.1109/TSG.2012.2223242.

56 Hu, R. and Qian, Y., Recent advances in communication infrastructures for smart grid, http://icc2014.ieee-icc.org/2014/private/Tutorial7.pdf. [Online; accessed 9-July-2017].

57 Nguyen, C. and Flueck, A. (2011) Modeling of communication latency in smart grid, in *Power and Energy Society General Meeting, 2011 IEEE*, pp. 1–7, doi:10.1109/PES.2011.6039815.

58 Kim, J., Kim, D., Lim, K.W., Ko, Y.B., and Lee, S.Y. (2012) Improving the reliability of ieee 802.11s based wireless mesh networks for smart grid systems. *Communications and Networks, Journal of*, **14** (6), 629–639, doi:10.1109/JCN.2012.00029.

59 Group, I.S..W. *et al.*, Ieee standard for local and metropolitan area networks.

60 (2008), Wireless Connectivity for Electric Substations, http://www.epri.com/abstracts/Pages/ProductAbstract.aspx? ProductId=000000000001016145. [Online; accessed 14-Sept-2017].

61 FCC Rules for Unlicensed Wireless Equipment operating in the ISM bands., http://www.afar.net/tutorials/fcc-rules. [Online; accessed 6-June-2015].

62 Xu, Y. and Wang, W. (2013) Wireless mesh network in smart grid: Modeling and analysis for time critical communications. *Wireless Communications, IEEE Transactions on*, **12** (7), 3360–3371, doi:10.1109/TWC.2013.061713.121545.

63 Gharavi, H. and Hu, B. (2011) Multigate communication network for smart grid. *Proceedings of the IEEE*, **99** (6), 1028–1045, doi:10.1109/JPROC.2011.2123851.

64 Al-Anbagi, I., Erol-Kantarci, M., and Mouftah, H. (2013) A reliable ieee 802.15.4 model for cyber physical power grid monitoring systems. *Emerging Topics in Computing, IEEE Transactions on*, **1** (2), 258–272, doi:10.1109/TETC.2013.2281192.

65 Rajalingham, G., Ho, Q.D., and Le-Ngoc, T. (2013) Attainable throughput, delay and scalability for geographic routing on smart grid neighbor area networks, in *Wireless Communications and Networking Conference (WCNC), 2013 IEEE*, pp. 1121–1126, doi:10.1109/WCNC.2013.6554721.

66 (2003), Communication networks and systems in substations– Part 5: Communication requirements for functions and device models, P-IEC 61850-5 ed1.0.

67 Chu, H.C., Siao, W.T., Wu, W.T., and Huang, S.C. (2011) Design and implementation an energy-aware routing mechanism for solar wireless sensor networks, in *High Performance Computing and Communications (HPCC), 2011 IEEE 13th International Conference on*, pp. 881–886, doi:10.1109/HPCC.2011.126.

68 Aksanli, B. and Rosing, T. (2013) Optimal battery configuration in a residential home with time-of-use pricing, in *Smart Grid Communications (SmartGridComm), 2013 IEEE International Conference on*, pp. 157–162, doi:10.1109/SmartGridComm.2013.6687950.

69 (2015), How to Prolong Lithium-based Batteries., http://batteryuniversity.com/learn/article/how_to_prolong_lithium_based_batteries. [Online; accessed 6-June-2015].

70 NIST Priority Action Plan 2, Guidelines for Assessing Wireless Standards for Smart Grid Applications., http://ftp.tiaonline.org/TR-45/TR-45.5/Incoming/TR-455-20110124__Seattle/500-11012403__NIST_PAP2_Guidelines_for_Assessing_Wireless_Stds_for_Smart_Grid_Appl_1.0.pdf. [Online; accessed 6-June-2015].

71 Markkula, J. and Haapola, J. (2013) Lte and hybrid sensor-lte network performances in smart grid demand response scenarios, in *Smart Grid Communications (SmartGrid-Comm), 2013 IEEE International Conference on*, pp. 187–192, doi:10.1109/SmartGridComm.2013.6687955.

72 Goodman, D. and Mandayam, N. (1999) Power control for wireless data, in *Mobile Multimedia Communications, 1999. (MoMuC '99) 1999 IEEE International Workshop on*, pp. 55–63, doi:10.1109/MOMUC.1999.819473.

73 Meshkati, F., Chiang, M., Poor, H., and Schwartz, S. (2006) A game-theoretic approach to energy-efficient power control in multicarrier cdma systems. *Selected Areas in Communications, IEEE Journal on*, **24** (6), 1115–1129, doi:10.1109/JSAC.2005.864028.

74 Lasaulce, S., Hayel, Y., El Azouzi, R., and Debbah, M. (2009) Introducing hierarchy in energy games. *Wireless Communications, IEEE Transactions on*, **8** (7), 3833–3843, doi:10.1109/TWC.2009.081443.

75 Meshkati, F., Poor, H., Schwartz, S., and Mandayam, N.B. (2005) An energy-efficient approach to power control and receiver design in wireless data networks. *Communications, IEEE Transactions on*, **53** (11), 1885–1894, doi:10.1109/TCOMM.2005.858695.

76 Rodriguez, V. (2003) An analytical foundation for resource management in wireless communication, in *Global Telecommunications Conference, 2003. GLOBECOM '03. IEEE*, vol. 2, pp. 898–902 Vol. 2, doi:10.1109/GLOCOM.2003.1258369.

77 Strassen algorithm, http://en.wikipedia.org/wiki/Strassen_algorithm.

78 Danahy, J. (2009), The Coming Smart Grid Data Surge, http://www.smartgridnews.com/story/coming-smart-grid-data-surge/2009-10-05. [Online; accessed 6-June-2015].

79 Hiertz, G.R., Denteneer, D., Max, S., Taori, R., Cardona, J., Berlemann, L., and Walke, B. (2010) IEEE 802.11 s: the WLAN mesh standard. *Wireless Communications, IEEE*, **17** (1), 104–111.

80 Gharavi, H. and Xu, C. (2012) Traffic scheduling technique for smart grid advanced metering applications. *Communications, IEEE Transactions on*, **60** (6), 1646–1658, doi:10.1109/TCOMM.2012.12.100620.

81 Miao, G., Himayat, N., Li, G., Koc, A., and Talwar, S. (2009) Interference-aware energy-efficient power optimization, in *Communications, 2009. ICC '09. IEEE International Conference on*, pp. 1–5, doi:10.1109/ICC.2009.5199096.

82 Kayastha, N., Niyato, D., Hossain, E., and Han, Z. (2014) Smart grid sensor data collection, communication, and networking: a tutorial. *Wireless communications and mobile computing*, **14** (11), 1055–1087.

83 Wang, Y., Saad, W., Han, Z., Poor, H., and Basar, T. (2014) A game-theoretic approach to energy trading in the smart grid. *Smart Grid, IEEE Transactions on*, **5** (3), 1439–1450, doi:10.1109/TSG.2013.2284664.

84 Ma, R., Chen, H.H., Huang, Y.R., and Meng, W. (2013) Smart grid communication: Its challenges and opportunities. *Smart Grid, IEEE Transactions on*, **4** (1), 36–46, doi:10.1109/TSG.2012.2225851.

85 Nishimura, F., Cicarelli, L., Arellano, R., and Soares, M. (2006) Opgw installation in energized transmission line, in *Transmission Distribution Conference and Exposition: Latin America, 2006. TDC '06. IEEE/PES*, pp. 1–8, doi:10.1109/TDCLA.2006.311398.

86 Ali, S., Alvi, B., and Asif, M. (2008) Opgw - our experience in kesc, in *Electric Power Conference, 2008. EPEC 2008. IEEE Canada*, pp. 1–6, doi:10.1109/EPC.2008.4763296.

87 Guo, Z., Ye, F., Guo, J., Liang, Y., Xu, G., Zhang, X., and Qian, Y. (2014) A wireless sensor network for monitoring smart grid transmission lines, in *Computer Communication and Networks (ICCCN), 2014 23rd International Conference on*, pp. 1–6, doi:10.1109/ICCCN.2014.6911790.

88 Cataliotti, A., Di Cara, D., Emanuel, A., and Nuccio, S. (2008) Characterization of current transformers in the presence of harmonic distortion, in *Instrumentation and Measurement Technology Conference Proceedings, 2008. IMTC 2008. IEEE*, pp. 2074–2078, doi:10.1109/IMTC.2008.4547390.

89 Cataliotti, A., Di Cara, D., Di Franco, P., Emanuel, A., and Nuccio, S. (2008) Frequency response of measurement current transformers, in *Instrumentation and Measurement Technology Conference Proceedings, 2008. IMTC 2008. IEEE*, pp. 1254–1258, doi:10.1109/IMTC.2008.4547234.

90 Lijia, R., Hong, L., and Yan, L. (2012) On-line monitoring and prediction for transmission line sag, in *Condition Monitoring and Diagnosis (CMD), 2012 International Conference on*, pp. 813–817, doi:10.1109/CMD.2012.6416272.

91 Sun, X., Lui, K., Wong, K., Lee, W., Hou, Y., Huang, Q., and Pong, P. (2011) Novel application of magnetoresistive sensors for high-voltage transmission-line monitoring. *Magnetics, IEEE Transactions on*, **47** (10), 2608–2611, doi:10.1109/TMAG.2011.2158085.

92 Huang, X. and Wei, X. (2012) A new on-line monitoring technology of transmission line conductor icing, in *Condition Monitoring and Diagnosis (CMD), 2012 International Conference on*, pp. 581–585, doi:10.1109/CMD.2012.6416211.

93 Xiao, S. (2009) Consideration of technology for constructing chinese smart grid. *Automation of Electric Power Systems*, **8** (1), 18–28.

94 Li, J. (2011) Analysis on strategic significance of construction and development of smart grid in china. *Journal of Changjiang Engineering Vocational College*, **8** (1), 18–28.

95 Zhang, K. and Li, H. (2011) The transmission strategy for energy harvesting wireless transmitters, in *Global Telecommunications Conference (GLOBECOM 2011), 2011 IEEE*, pp. 1–5, doi:10.1109/GLOCOM.2011.6134169.

96 Bu, S., Yu, F., Cai, Y., and Liu, X. (2012) When the smart grid meets energy-efficient communications: Green wireless cellular networks powered by the smart grid. *Wireless Communications, IEEE Transactions on*, **11** (8), 3014–3024, doi:10.1109/TWC.2012.052512.111766.

97 Lin, J., Zhu, B., Zeng, P., Liang, W., Yu, H., and Xiao, Y. (2014) Monitoring power transmission lines using a wireless sensor network. *Wirel. Commun. Mob. Comput.*, doi:10.1002/wcm.2458.

98 Wijaya, T., Larson, K., and Aberer, K. (2013) Matching demand with supply in the smart grid using agent-based multiunit auction, in *Communication Systems and Networks (COMSNETS), 2013 Fifth International Conference on*, pp. 1–6, doi:10.1109/COMSNETS.2013.6465595.

99 Mohsenian-Rad, A.H., Wong, V., Jatskevich, J., Schober, R., and Leon-Garcia, A. (2010) Autonomous demand-side management based on game-theoretic energy consumption scheduling for the future smart grid. *Smart Grid, IEEE Transactions on,* 1 (3), 320–331, doi:10.1109/TSG.2010.2089069.

100 Bu, S. and Yu, F. (2013) A game-theoretical scheme in the smart grid with demand-side management: Towards a smart cyber-physical power infrastructure. *Emerging Topics in Computing, IEEE Transactions on,* 1 (1), 22–32, doi:10.1109/TETC.2013.2273457.

101 Zhu, Q., Han, Z., and Basar, T. (2012) A differential game approach to distributed demand side management in smart grid, in *Communications (ICC), 2012 IEEE International Conference on,* pp. 3345–3350, doi:10.1109/ICC.2012.6364562.

102 Salehfar, H. and Patton, A. (1989) Modeling and evaluation of the system reliability effects of direct load control. *Power Systems, IEEE Transactions on,* 4 (3), 1024–1030, doi:10.1109/59.32594.

103 Kondoh, J. (2011) Direct load control for wind power integration, in *Power and Energy Society General Meeting, 2011 IEEE,* pp. 1–8, doi:10.1109/PES.2011.6039480.

104 Nguyen, H.K., Song, J., and Han, Z. (2012) Demand side management to reduce peak-to-average ratio using game theory in smart grid, in *Computer Communications Workshops (INFOCOM WKSHPS), 2012 IEEE Conference on,* pp. 91–96, doi:10.1109/INFCOMW.2012.6193526.

105 Ramachandran, B., Srivastava, S.K., Edrington, C.S., and Cartes, D.A. (2011) An intelligent auction scheme for smart grid market using a hybrid immune algorithm. *Industrial Electronics, IEEE Transactions on,* 58 (10), 4603–4612.

106 Li, D., Jayaweera, S.K., and Naseri, A. (2011) Auctioning game based demand response scheduling in smart grid, in *Online Conference on Green Communications (GreenCom), 2011 IEEE,* IEEE, pp. 58–63.

107 (2014), Retail Sales of Electricity to Ultimate Customers, http://www.eia.gov/electricity/monthly/epm_table_grapher.cfm?t=epmt_5_01. [Online; accessed 6-June-2015].

108 (2014), Net Generation by Energy Source: Total (All Sectors), 2003 - 2013, http://www.eia.gov/electricity/annual/html/epa_03_01_a.html. [Online; accessed 6-June-2015].

109 (2014), Renewable Energy Cost Database, http://www.epa.gov/
cleanenergy/energy-resources/renewabledatabase.html. [Online;
accessed 6-June-2015].

110 Your guide to renewable energy, http://www.epa.gov/cleanenergy/
energy-resources/renewabledatabase.html. [Online; accessed
6-June-2015].

111 Carbon tax center, http://www.carbontax.org. [Online; accessed
6-June-2015].

112 (2014), Levelized Cost and Levelized Avoided Cost of New Generation Resources in the Annual Energy Outlook 2014, http://
www.eia.gov/forecasts/aeo/electricity_generation.cfm. [Online;
accessed 6-June-2015].

113 Stephen Boyd, L.V. (2004) *Convex Optimization*, Cambridge
University Press.

114 Rosen, J.B. (1965) Existence and uniqueness of equilibrium points
for concave n-person games. *Econometrica: Journal of the Econo-
metric Society*, pp. 520–534.

115 Bahrami, S. and Parniani, M. (2014) Game theoretic based charging strategy for plug-in hybrid electric vehicles. *Smart Grid, IEEE
Transactions on*, **5** (5), 2368–2375, doi:10.1109/TSG.2014.2317523.

116 Feng, B., Yu, R., and Lai, Y. (2013) Efficient and fair scheduling of phev charging with neuro dynamic programming, in
*Communications and Networking in China (CHINACOM),
2013 8th International ICST Conference on*, pp. 930–935,
doi:10.1109/ChinaCom.2013.6694728.

117 Yang, Z., Xu, W., and Yu, X. (2013) Optimal phev charge
scheduling for additional power loss ratio and charging
cost minimizations, in *Electrical Machines and Systems
(ICEMS), 2013 International Conference on*, pp. 465–469,
doi:10.1109/ICEMS.2013.6754568.

118 Ye, F., Qian, Y., and Hu, R.Q. (2017) Incentive load scheduling
schemes for phev battery exchange stations in smart grid. *IEEE
Systems Journal*, **11** (2), 922–930.

119 Huang, J., Gupta, V., and Huang, Y.F. (2012) Scheduling
algorithms for phev charging in shared parking lots, in
American Control Conference (ACC), 2012, pp. 276–281,
doi:10.1109/ACC.2012.6314939.

120 Wu, C., Mohsenian-Rad, H., and Huang, J. (2012)
Vehicle-to-aggregator interaction game. *Smart Grid, IEEE Transactions on*, **3** (1), 434–442, doi:10.1109/TSG.2011.2166414.

121 Liu, C., Wang, J., Botterud, A., Zhou, Y., and Vyas, A. (2012) Assessment of impacts of phev charging patterns on wind-thermal scheduling by stochastic unit commitment. *Smart Grid, IEEE Transactions on*, **3** (2), 675–683, doi:10.1109/TSG.2012.2187687.

122 Moeini-Aghtaie, M., Abbaspour, A., and Fotuhi-Firuzabad, M. (2014) Online multicriteria framework for charging management of phevs. *Vehicular Technology, IEEE Transactions on*, **63** (7), 3028–3037, doi:10.1109/TVT.2014.2320963.

123 Dong, Q., Niyato, D., Wang, P., and Han, Z. (2013) An adaptive scheduling of phev charging: Analysis under imperfect data communication, in *Smart Grid Communications (SmartGridComm), 2013 IEEE International Conference on*, pp. 205–210, doi:10.1109/SmartGridComm.2013.6687958.

124 Zhou, K. and Cai, L. (2014) Randomized phev charging under distribution grid constraints. *Smart Grid, IEEE Transactions on*, **5** (2), 879–887, doi:10.1109/TSG.2013.2293733.

125 Scacchioli, A., Rizzoni, G., Salman, M., Li, W., Onori, S., and Zhang, X. (2014) Model-based diagnosis of an automotive electric power generation and storage system. *Systems, Man, and Cybernetics: Systems, IEEE Transactions on*, **44** (1), 72–85, doi:10.1109/TSMCC.2012.2235951.

126 Yukita, K., Ichiyanagi, K., Goto, Y., and Hirose, K. (2007) A study of electric power quality using storage system in distributed generation, in *Electrical Power Quality and Utilisation, 2007. EPQU 2007. 9th International Conference on*, pp. 1–4, doi:10.1109/EPQU.2007.4424142.

127 Shao, S., Pipattanasomporn, M., and Rahman, S. (2012) Grid integration of electric vehicles and demand response with customer choice. *Smart Grid, IEEE Transactions on*, **3** (1), 543–550, doi:10.1109/TSG.2011.2164949.

128 Lee, T.K., Bareket, Z., Gordon, T., and Filipi, Z. (2012) Stochastic modeling for studies of real-world phev usage: Driving schedule and daily temporal distributions. *Vehicular Technology, IEEE Transactions on*, **61** (4), 1493–1502, doi:10.1109/TVT.2011.2181191.

129 Wikipedia, Big data, https://en.wikipedia.org/wiki/Big_data. [Online; accessed 23-July-2017].

130 Hilbert, M. and López, P. (2011) The world?s technological capacity to store, communicate, and compute information. *science*, **332** (6025), 60–65.

131 Kezunovic, M. (2017), Big data applications in smart grids: benefits and challenges, IEEE Smartgrid.

132 Mahmoud Daneshman, K.J.L. (2017), Big challenges for big data in the smart grid era, https://www.ecnmag.com/blog/2017/04/big-challenges-big-data-smart-grid-era. [Online; accessed 23-July-2017].

133 Simmhan, Y., Aman, S., Kumbhare, A., Liu, R., Stevens, S., Zhou, Q., and Prasanna, V. (2013) Cloud-based software platform for big data analytics in smart grids. *Computing in Science Engineering*, **15** (4), 38–47, doi:10.1109/MCSE.2013.39.

134 Cheung, J., Czaszejko, T., and Morton, A. (2007) Transmission loss evaluation in an open electricity market using an incremental method. *Generation, Transmission Distribution, IET*, **1** (1), 189–196, doi:10.1049/iet-gtd:20050332.

135 Aman, S., Simmhan, Y., and Prasanna, V. (2011) Improving energy use forecast for campus micro-grids using indirect indicators, in *Data Mining Workshops (ICDMW), 2011 IEEE 11th International Conference on*, pp. 389–397, doi:10.1109/ICDMW.2011.95.

136 Tan, W., Blake, M., Saleh, I., and Dustdar, S. (2013) Social-network-sourced big data analytics. *Internet Computing, IEEE*, **17** (5), 62–69, doi:10.1109/MIC.2013.100.

137 Hu, H., Wen, Y., Chua, T.S., and Li, X. (2014) Toward scalable systems for big data analytics: A technology tutorial. *Access, IEEE*, **2**, 652–687, doi:10.1109/ACCESS.2014.2332453.

138 (2013), Smart data set for sustainability, http://traces.cs.umass.edu/index.php/Smart/Smart. [Online; accessed 6-July-2015].

139 Barker, S., Mishra, A., Irwin, D., Cecchet, E., Shenoy, P., and Albrecht, J. (2012) Smart★: An open data set and tools for enabling research in sustainable homes. *SustKDD, August*.

140 Ford, V., Siraj, A., and Eberle, W. (2014) Smart grid energy fraud detection using artificial neural networks, in *Computational Intelligence Applications in Smart Grid (CIASG), 2014 IEEE Symposium on*, pp. 1–6, doi:10.1109/CIASG.2014.7011557.

141 Liu, Y., Ning, P., and Reiter, M.K. (2011) False data injection attacks against state estimation in electric power grids. *ACM Transactions on Information and System Security (TISSEC)*, **14** (1), 13.

142 Ozay, M., Esnaola, I., Yarman Vural, F.T., Kulkarni, S.R., and Poor, H.V. (2015) Machine learning methods for attack detection in the smart grid.

143 Fabris, F., Margoto, L., and Varejao, F. (2009) Novel approaches
     for detecting frauds in energy consumption, in *Network and Sys-
     tem Security, 2009. NSS '09. Third International Conference on,*
     pp. 546–551, doi:10.1109/NSS.2009.17.
144 (2011), The nist definition of cloud computing, http://csrc.nist
     .gov/publications/nistpubs/800-145/SP800-145.pdf. [Online;
     accessed 6-June-2015].
145 Silverstone, R. and Haddon, L. (1996) Design and the domestica-
     tion of information and communication technologies: Technical
     change and everyday life.
146 Chen, M., Mao, S., and Liu, Y. (2014) Big data: A survey. *Mobile
     Networks and Applications,* **19** (2), 171–209.
147 Xu, S., Qian, Y., and Hu, R.Q. (2016) A secure data learning
     scheme in big data applications, in *Computer Communication and
     Networks (ICCCN), 2016 25th International Conference on,* IEEE,
     pp. 1–9.
148 Xu, L., Jiang, C., Wang, J., Yuan, J., and Ren, Y. (2014) Informa-
     tion security in big data: privacy and data mining. *IEEE Access,* **2**,
     1149–1176.
149 Du, W., Han, Y.S., and Chen, S. (2004) Privacy-preserving multi-
     variate statistical analysis: Linear regression and classification, in
     *Proceedings of the 2004 SIAM international conference on data
     mining,* SIAM, pp. 222–233.
150 Shokri, R. and Shmatikov, V. (2015) Privacy-preserving deep
     learning, in *Proceedings of the 22nd ACM SIGSAC conference on
     computer and communications security,* ACM, pp. 1310–1321.
151 Lindell, Y. and Pinkas, B. (2009) Secure multiparty computa-
     tion for privacy-preserving data mining. *Journal of Privacy and
     Confidentiality,* **1** (1), 5.
152 Clifton, C., Kantarcioglu, M., Vaidya, J., Lin, X., and Zhu, M.Y.
     (2002) Tools for privacy preserving distributed data mining. *ACM
     Sigkdd Explorations Newsletter,* **4** (2), 28–34.
153 Brickell, J. and Shmatikov, V. (2009) Privacy-preserving classifier
     learning., in *Financial Cryptography,* Springer, pp. 128–147.
154 Vaidya, J., Shafiq, B., Basu, A., and Hong, Y. (2013) Differen-
     tially private naive bayes classification, in *Proceedings of the 2013
     IEEE/WIC/ACM International Joint Conferences on Web Intelli-
     gence (WI) and Intelligent Agent Technologies (IAT)-Volume 01,*
     IEEE Computer Society, pp. 571–576.

155 Vapnik, V. (2013) *The nature of statistical learning theory*, Springer science & business media.

156 Bishop, C.M. (2006) *Pattern recognition and machine learning*, springer.

157 Boyd, S. and Vandenberghe, L. (2004) *Convex optimization*, Cambridge university press.

158 James, G., Witten, D., Hastie, T., and Tibshirani, R. (2013) *An introduction to statistical learning*, vol. 112, Springer.

159 Ng, A.Y. (2004) Feature selection, l 1 vs. l 2 regularization, and rotational invariance, in *Proceedings of the twenty-first international conference on Machine learning*, ACM, p. 78.

160 Bottou, L. (2010) Large-scale machine learning with stochastic gradient descent, in *Proceedings of COMPSTAT'2010*, Springer, pp. 177–186.

161 Chaum, D., Crépeau, C., and Damgard, I. (1988) Multiparty unconditionally secure protocols, in *Proceedings of the twentieth annual ACM symposium on Theory of computing*, ACM, pp. 11–19.

162 Blum, M., Feldman, P., and Micali, S. (1988) Non-interactive zero-knowledge and its applications, in *Proceedings of the twentieth annual ACM symposium on Theory of computing*, ACM, pp. 103–112.

163 Repository, Smart-UMass Trace Repository, http://traces.cs.umass .edu/. [Online; accessed 11-5-2017].

164 Ye, F. and Qian, Y. (2015) Big data and cloud computing based demand-side management for electric vehicles in smart grid. *E-LETTER*.

165 Xu, S. and Qian, Y. (2015) Quantitative study of reliable communication infrastructure in smart grid nan, in *Design of Reliable Communication Networks (DRCN), 2015 11th International Conference on the*, IEEE, pp. 93–94.

166 Xu, S., Qian, Y., and Hu, R.Q. (2015) On reliability of smart grid neighborhood area networks. *IEEE Access*, **3**, 2352–2365.

167 Ye, F. and Qian, Y. (2015) Secure communication networks in the advanced metering infrastructure of smart grid. *ZTE Commun.*, **13** (3), 13–20.

168 Xu, S., Qian, Y., and Hu, R.Q. (2017) A data-driven preprocessing scheme on anomaly detection in big data applications, in *IEEE INFOCOM 2017 BigSecurity Workshop*, IEEE.

169 Ye, F., Qian, Y., and Hu, R.Q. (2015) Energy efficient self-sustaining wireless neighborhood area network design for smart grid. *IEEE Transactions on Smart Grid*, **6** (1), 220–229.

170 Xu, S., Qian, Y., and Hu, R.Q. (2017) A study on communication network reliability for advanced metering infrastructure in smart grid, in *2017 IEEE Cyber Science and Technology Congress (CyberSciTech 2017)*, IEEE.

171 Ye, F., Qian, Y., and Hu, R. (2014) A security protocol for advanced metering infrastructure in smart grid, in *Global Communications Conference (GLOBECOM), 2014 IEEE*, pp. 649–654, doi:10.1109/GLOCOM.2014.7036881.

172 Guidelines for smart grid cyber security., http://dx.doi.org/10.6028/NIST.IR.7628r1. [Online; accessed 23-July-2017].

173 Wikipedia, Stuxnet, https://en.wikipedia.org/wiki/Stuxnet. [Online; accessed 21-July-2017].

174 Morris, T.H., Pan, S., and Adhikari, U. (2012) Cyber security recommendations for wide area monitoring, protection, and control systems, in *2012 IEEE Power and Energy Society General Meeting*, pp. 1–6, doi:10.1109/PESGM.2012.6345127.

175 Zhang, Z., Gong, S., Dimitrovski, A.D., and Li, H. (2013) Time synchronization attack in smart grid: Impact and analysis. *IEEE Transactions on Smart Grid*, **4** (1), 87–98, doi:10.1109/TSG.2012.2227342.

176 Jokar, P., Nicanfar, H., and Leung, V.C.M. (2011) Specification-based intrusion detection for home area networks in smart grids, in *2011 IEEE International Conference on Smart Grid Communications (SmartGridComm)*, pp. 208–213, doi:10.1109/SmartGridComm.2011.6102320.

177 Hu, B. and Gharavi, H. (2014) Smart grid mesh network security using dynamic key distribution with merkle tree 4-way handshaking. *IEEE Transactions on Smart Grid*, **5** (2), 550–558, doi:10.1109/TSG.2013.2277963.

178 Lin, H.Y., Shen, S.T., and Lin, B.S.P. (2012) A privacy preserving smart metering system supporting multiple time granularities, in *2012 IEEE Sixth International Conference on Software Security and Reliability Companion*, pp. 119–126, doi:10.1109/SERE-C.2012.22.

179 Finster, S. (2013) Smart meter speed dating, short-term relationships for improved privacy in smart metering, in *2013 IEEE International Conference on Smart Grid Communications (SmartGridComm)*, pp. 426–431, doi:10.1109/SmartGridComm.2013.6687995.

180 Ukil, A. and Zivanovic, R. (2014) Automated analysis of power systems disturbance records: Smart grid big data perspective, in *Innovative Smart Grid Technologies - Asia (ISGT Asia), 2014 IEEE*, pp. 126–131, doi:10.1109/ISGT-Asia.2014.6873776.

181 Bitzer, B. and Gebretsadik, E. (2013) Cloud computing framework for smart grid applications, in *Power Engineering Conference (UPEC), 2013 48th International Universities'*, pp. 1–5, doi:10.1109/UPEC.2013.6714855.

182 Baek, J., Vu, Q., Liu, J., Huang, X., and Xiang, Y. (2014 (early access)) A secure cloud computing based framework for big data information management of smart grid. *Cloud Computing, IEEE Transactions on*, pp. 1–1, doi:10.1109/TCC.2014.2359460.

183 Waters, B. (2005) Efficient identity-based encryption without random oracles, in *Advances in Cryptology–EUROCRYPT 2005*, Springer, pp. 114–127.

184 Libert, B. and Quisquater, J.J. (2004) The exact security of an identity based signature and its applications. *IACR Cryptology ePrint Archive*, **2004**, 102.

185 Ye, F., Qian, Y., and Hu, R.Q. (2015) HIBaSS: hierarchical identity-based signature scheme for AMI downlink transmission. *Security and Communication Networks*, **8** (16), 2901–2908.

186 IEEE Std 2030 (2011), Guide for Smart Grid Interoperability of Energy Technology and Information Technology Operation with the Electric Power System (EPS), and End-Use Applications and Loads, IEEE Standards Association.

187 Burnett, S. and Paine, S. (2001) *The RSA Security's Official Guide to Cryptography*, McGraw-Hill, Inc.

188 Boneh, D. and Franklin, M. (2003) Identity-based encryption from the weil pairing. *SIAM Journal on Computing*, **32** (3), 586–615.

189 Rusitschka, S., Eger, K., and Gerdes, C. (2010) Smart grid data cloud: A model for utilizing cloud computing in the smart grid domain, in *Smart Grid Communications (SmartGridComm), 2010 First IEEE International Conference on*, pp. 483–488, doi:10.1109/SMARTGRID.2010.5622089.

190 Zheng, L., Hu, Y., and Yang, C. (2011) Design and research on private cloud computing architecture to support smart grid, in *Intelligent Human-Machine Systems and Cybernetics (IHMSC), 2011 International Conference on*, vol. 2, vol. 2, pp. 159–161, doi:10.1109/IHMSC.2011.109.

191 Park, C.M., Kim, M.H., and Yung, M. (2005) A remark on implementing the weil pairing, in *Information Security and Cryptology*, Springer, pp. 313–323.

192 Joux, A. and Nguyen, K. (2003) Separating decision diffie–hellman from computational diffie–hellman in cryptographic groups. *Journal of cryptology*, **16** (4), 239–247.

Index